SENSORIUM 458

The Myriad Masks of Reality

THOMAS CHARRINGTON

Copyright © 2021 Thomas Charrington

The right of Thomas Charrington to be identified as the Author of the Work, has been asserted by him in accordance with the Copyright, Designs and Patents Act 1988.

All rights reserved. No part of this publication may be reproduced, stored in a retrieval system, or transmitted, in any form or by any means without the prior written permission of the author or his agents.

All characters and commercial enterprises in this publication are fictitious (excluding factual historical references) and any resemblance to real enterprises and real persons, living or dead, is purely coincidental.

Cover Design © Estignotum

Preface

I must thank you for choosing this book. It suggests you have an interest in a topic that has shadowed me for decades. Although the subject matter is quite profound, many of you will be put off by the fact that it is not presented in a serious academic format. Instead it is written as a dialogue between two young men who use the sort of language that young people would use in casual conversation. I decided that the argumentative 'format' (or dialectical method) would make it easier to grasp the startling truths embedded in these pages. Truths which are still oddly elusive and remain veiled, even in the modern world – perhaps more so.

I am not an academic. I don't have a degree in philosophy or psychology or neuroscience or parapsychology. I get that many people will immediately dismiss the utterings of an 'uneducated' layman as without any true and proper authenticity. In the modern age, only those with university and post graduate accolades are to be trusted in the analysis and dissemination of 'serious' information. This is why, when John E Mack wrote a book called 'Abduction' (a subject that would normally have the academic fraternity sniggering into its shirt sleeves) a great fuss was made of the fact that he was a 'Harvard professor' and therefore his writings must be imbued with some sort of gravitas. In other words, his credibility rested on his academic prowess.

In fields such as engineering, architecture, mathematics, it is of course imperative that a high degree of study has been achieved prior to engaging in these disciplines. Personally however, I do not think this is true of philosophical 'thinking' or speculation in relation to human perception.

My mentors are mostly dead, but their thought processes live on in their writings. Ouspensky, Gurdjieff, Nietzsche, Kant, Schopenhauer, Berkeley, Castaneda to name but a few, still speak their wisdom across the vaults of time. Most of them allude to the hidden truth that has been obscured by the brilliance of our technological advances. A truth that won't go away and which, unless addressed, will keep Homo sapiens forever in the dark. Human consciousness is a subtle subject and we are by no means at a point of finality as to what it actually is and indeed what possibilities lie within its 'guided' manipulation and control.

Prologue

A school master stood in a classroom before several rows of bored looking children. On a table in front of him, a life size plastic human brain was perched on a plinth.

"This," he said, tapping the convoluted lobes lightly with his pointer, whilst sweeping his eyes around the room, "is the human brain, which as you can see looks not unlike a walnut. This organ is possibly the most complex thing in the universe - its cognitive capabilities are stupendous." The children stared at the brain dreamily and then back at the teacher.

"This small object and countless others like it, working in unison," he continued, "have put the human race where it is today ….at the top of the evolutionary ladder and able to manipulate this planet and probably many more to come, in ways which leave us way out in front. No other creature has anything like our intelligence. We are unique. The world, (he swept his pointer in a circle around his head) is analysed by this (he tapped the brain) and as a species we move forward."

But there was something, the teacher hadn't quite grasped. The 'brain' wasn't analysing the classroom in which he and his students were present that Tuesday morning - it had already done that.

No ...the place they sat in was already a mind construct, and both he, his students and the brain on the plinth, were all a part of it - trapped within its walls, slaves to its bidding and blind to its machinations.

Oh man! Attend!
What does deep midnight's voice contend?
I slept my sleep,
And now awake at dreaming's end:
The world is deep,
And deeper than day can comprehend…

Friedrich Nietzsche
(Clipped)

Sensorium 458 – A Dialogue

Adam yanked the wheel of his old car and moved onto the rough ground at the side of the Military Road. Dry chalky dust rolled forwards over the scarred bodywork like smoke, as he stepped out into the brilliant sunshine. He stretched whilst Peter fiddled with his camera in the passenger seat.

"We'll have a spot of lunch over there by that abandoned cottage," he said, having scanned the surroundings. "It'll give us a wonderful view to the southwest and you can tell me what you've been learning at Cambridge over the passed couple of years."

"Hey, it's really not so exciting. Field work is what I crave, not listening to jaded professors in lecture theatres," Peter replied, putting his camera back in its case. He got out and shouldered his rucksack. "God, I've missed the Isle of Wight!" he said, taking a deep breath of salty air, "It's got such a history … human and reptilian!"

"True. There've been some big beasts in these parts," Adam said, marching ahead, "and I'm sure you'll find something to take home. Something a little more interesting than a mere ammonite. Perhaps a tyrannosaur tooth! C'mon, I'm getting hungry!"

They wandered through a long stretch of coarse grass towards the cliff edge and became aware of the distant roar of surf. After laying down a rug they began chewing through the various containers of food, Adam's wife had cobbled together earlier. They were old friends and although they hadn't seen much of each other over the passed few years, conversation was easy and relaxed. They chatted lazily for an hour or so and then Adam suddenly sat up and wrapped his arms around his knees. He stared into the distance for a time, in complete silence whilst Peter teased an ant with a grass stem.

"It's one of those endless battles, a view like this," he said eventually, in a low calm voice.

Peter lifted his head off the rug. "I'm sorry?" he said.

"An endless battle between what one knows to be true and the unconquerable impression to the opposite," Adam continued, still staring out.

"I'm really not with you, Adam?"

"Oh, of course you're not," Adam said, turning to look at Peter, "Why would you be. What I mean is ... the world seems so 'out there'. Just look at that huge expanse of sea and sky spreading out to the horizon …but we both know that it is in fact …well, mind …consciousness…perception. It's 'out thereness' is just an illusion."

"Uh? Please explain," Peter said with a wrinkled forehead, "you're surely not suggesting the world out there doesn't exist?"

Adam smiled and remained quiet for a few moments as though gathering his thoughts.

"Something exists out there my friend but it sure as hell is not the thing we see. Every day of our lives, from the moment we're born to the moment we die, we imagine that the outside world is …well …outside of us, when the simple truth is that it is not. It is our mind we live in, Peter, not the outside world. Frankly, it has to be one of the weirdest realisations that any human being can have."

"Bit heavy for a glorious afternoon my old friend!" Peter said, sitting up. "The world impresses itself on us and we respond accordingly. We're equipped with five senses to navigate our world - otherwise we'd be doing silly things like walking off cliffs and drinking sea water!"

"But that's just it!" Adam said suddenly more alert, "we're equipped to navigate 'the world' with our five senses. But Peter, we're continuously making an assumption that the world as we perceive it, is a fairly accurate description of 'the real deal' …that our eyes are like windows onto a real world… yes?"

"The real deal? I suppose what you mean is this?" Peter said, sweeping his arm around him. "In its crudist form I assume the world has a surface which I walk on, and that there are hard obstacles I need to be aware of or I'll break my bloody leg! Look Adam, I've come here to hunt for fossils and hopefully find a piece of dinosaur and we seem to be a bit sidelined."

"Hang on there my friend, this is an important topic and we have bags of time. Besides, the glare of the sun at this time makes those bones much more difficult to see. So, the world has a surface."

"Well, as we both know, a surface is merely a perimeter between two different densities of matter; you know …where the packing of the molecules or atoms suddenly changes."

"Mmm …of course," Adam said stroking his chin. "Now let's go back to what I was saying a few moments ago about our minds and the world. You see I have a problem with your idea - which is of course the same assumption made by almost every rational person on the planet."

"So you're happy there is a planet?" Peter interrupted with a grin.

"Ha ha. You see I've come to a very different conclusion and it goes like this. As anyone who's spent enough time thinking about it has realised, the world is pure mind. Do you understand this?"

"Go on."

"We're not in *the* world Peter …we're actually in *our* world. The world we live in, is only a model …the human representation of a world we cannot directly reach. But of course we're all so used to this world as an objective hard reality that exists outside of us, that such an idea is preposterous!"

"Too bloody right mate! I'm guessing what you mean is that the world is composed of tiny balls of matter that give the illusion of the objects we see and feel."

"Well not exactly. What I'm actually getting at is that our whole reality is self generated, including the atoms that form matter."

"Self generated?"

"Generated by our minds. And you and I are in that place right now. It's called the material world."

"This smacks of the 'Matrix' …have you been overdosing on sci- fi, mate?"

"Good film …but no, I don't think we're all being fed a synthetic reality by a master race of machines!"

"Okaaaay."

"Anyway let's stay on track, I want to get this off my chest. The world we humans inhabit, is in reality, only a language …not a place… or a collection of objects, in the way we understand it. It's the result of an interaction, if you like, between us and our environment."

"Adam, you're coming across as a bit cuckoo this afternoon."

"Perhaps I am, but that's only because you're unfamiliar with it. There's plenty in the world you inhabit, and that science holds true, that's cuckoo. Like the fact that the universe is supposed to have begun as something small enough to fit under your finger nail!"

"Very true!"

"What I'm saying about the world being a language, rather than a real 'place' is so obvious … and yet so impossible to stomach. It's not something you can learn, like how far it is to the moon … it's something that has to break onto you like a wave…a wave which imparts an understanding before rolling on, and leaving the normal world to stamp its authority on you once again."

Peter looked at him askance and clicked his tongue. "You're a weird one Adam you really are. What the hell are you talking about? The 'actual' world? What on earth is that?"

"The world or shall we say reality, before it's translated into our mind language."

"Mind language? I'm not with you. What exactly do you mean? How can the world be a language? Just sounds like a load of pretentious bullshit to me! Are you telling me, this is part of that language?" he said, holding up a small rock.

"Absolutely Peter…but that's a bit misleading. That's like a man who's dreaming he's walking through a forest, stopping, pointing at a tree and saying, 'that's a dream tree!' Everything you perceive is made up of our mind language – that rock, and the hand that holds it, and the sound of that truck that just went by and even the space that surrounds us. We live in mind Peter. You're only confused because all your life you've been under the impression that the outside world as we perceive it (he pointed with his chin to the horizon), is outside us!"

"Sure Adam …just keep thinking that!" Peter replied light heartedly.

"Do you really think this view, the feel of the breeze on your face, the sound of those larks, are features of the real world? Well they are not!" Adam continued, "they're only symbols. There is no *solid* world Peter, there is only energy."

"Sounds bloody crazy to me… and I know exactly where this is coming from."

"Really?"

"Sure, it's coming straight out of a little philosophy book you've been reading, by a fellow called Emmanuel Kant! Yeah?"

"No, you're wrong. I haven't *just* bought a book by Kant or anyone else for that matter though I have read them in the past."

"Aha! So I'm not so far wrong am I?" Peter chuckled.

"Look, Peter let's not worry about the where's and why's, and whether I just bought this or that book … which by the way isn't true anyway…let's just have an interesting discussion."

"Okay …why not. So you just said the outside world is in actual fact a mind language, yes?" Peter said crunching on a stick of celery, "well for a start you're not the first person to say that Adam. It's an old idea which has been kicking around for time immemorial, and … well, it just doesn't seem to ring true in our times".

"That's simply because we interact with this 'world' if you like …in an increasingly efficient manner. It seems to yield to our will!"

"Damn right! And the industrial revolution was just the beginning! Look at us now!" Peter said, chucking the stone towards the cottage."

"I know…but what I'm trying to get at is that …although the human race is moving ahead with break neck speed in a technological sense, we're still trapped within ourselves! We're dealing with 'mind' all the time and mistaking it for reality! Don't you see? We're stuck inside the workings of a super efficient navigation system. In effect our 'minds'!"

"Sure we are!"

"Listen, I don't expect you to say, 'Ok I get it'."

"Get it? Like you're some professor of philosophy, Adam!"

"Some information has to seep into you like osmosis. I can pressure you to tell me what sixty-nine times fifty-eight is, because the sharper your focus the better you can make the calculation - but this is the opposite. You have to reduce your focus."

"And act all goofy!" Peter interrupted

"It would be like trying to order you to go to sleep. Look, don't get angry, just listen to my logic for a few minutes. So, the hardness of our world is built by our minds - in effect, that perception we take to signify the existence of our trusty solid world …you know …the streets, the buildings, the seas, turns out to be our creation, our sensation, our reaction …"

"Our reaction to a very real and external world," Peter interjected.

"And if I can just finish….and not something that exists outside us at all ..*in the manner it presents itself.*"

"Well hang on, I just said that the world is made up of tiny particles of matter, so of course it isn't real in that sense. We just see lumps of matter. We're not designed to see it at the sub microscopic level. So I don't really see what you're driving at!"

"I'm not meaning that Peter. I'm not saying that matter is different because it's made up of tiny components. Like looking at a sandy beach from the top of a tree where the grains of sand are invisible. I'm saying that the whole concept of matter is made of mind."

"Adam, you're surely not suggesting that matter doesn't exist? Objects, things, hardness *are* external to us. You know that! Go and kick that rock and remind yourself!"

"No Peter, our experience of hardness … which if you think about it is what hardness actually is … is *not* outside us at all ... it *is* us! And as our experience or perception of hardness is *all* we've ever known, then it follows that we do indeed live in 'mind'. The rest is assumption or presumption or deduction."

"Have you heard of the periodic table, my friend?" Peter said sarcastically, "where matter has been comprehensively catalogued and its properties analysed to the point where there is no room for error?"

"Meaning?" Adam said puzzled.

"Meaning …meaning there is no doubt about its status as externally existing 'stuff' with qualities and characteristics that we've come to understand."

"Because matter has been analysed and its properties established and the way it interacts with other matter thoroughly examined and understood, does not bring it one step closer to it being something existing independently of your mind," Adam continued.

"Well how in the hell can we predict the course of chemical reactions if these are all made of mind and not externally existing matter then?" Peter said hotly.

"Precisely because we have learned the properties of these materials and how they interact," Adam replied.

"Exactly! Their objective properties. They exist independently of us Adam. They are part of our solid external environment!"

"Yes … they do exist independently of us, but in a condition quite unknown to us."

"But you've just contradicted yourself! You just said, we've learned their properties and how they interact…..so how can they exist in an unknown condition?"

"Because the materials or chemicals you are referring to, and which we are familiar with, are features of your perception ... they're made of mind. They are not in their independent condition. Don't you understand the subjective element of these observations? Do you really believe that the way in which these reactions display themselves to us have nothing to do with our perspective ... our cognition, our methodology? Peter, all these things are imbued with our 'mind tool' if you will, and its mechanisms, but we have no way of proving this because we're embedded within it."

"For Christ's sake Adam stop being an idiot! You cannot be suggesting that water won't boil into steam if I'm not there to watch it on the stove!"

"Like if you walked out of the room the whole process of boiling would stop?" Adam said with a humorous glint in his eye.

"No ...well actually yes!"

"Of course it wouldn't stop. In fact the pot would be nicely dry when you returned!"

"Look, I'm being serious here. What are you driving at? I refuse to believe that these objective things are somehow changing or not existing when I'm not here to see them."

"Well, my old friend, perhaps you had better start believing. Our reality is under the dictatorship of our subjective cognition. It pervades everything we experience. Everything, *like everything,* is under its command. From the views of the molecular world under an electron microscope to the vistas of space displayed by our most powerful telescopes. None of these things belong to some external, solid, enduring reality in the manner they present themselves! These are mind sculptures Peter. From solid granite to the empty vacuum of space ..built by our minds ... our perceptual machinery!"

"God help me," Peter muttered, "my buddy's gone nutty!"

"You see, we've been born into this world and ever since we began to breath, it has pretended most convincingly to be external to us. We run across the garden and bang into a tree and bruise our head - the garden, the grass, the tree, are all features of the outside world and the bruised head is us. But Peter, in truth they are *all* mind constructions. Yes there is an external reality, and yes we are living entities surviving within that reality. But the reality we are dealing with …living in, is 'perception'. It always has been! It's called the 'World'. And we are part of that world … a world made of perception."

"Come on Adam!"

"Don't take this the wrong way, but I don't think you fully understand what I'm saying.

You still haven't truly realised that absolutely everything you attribute to the 'external world' is actually a mind creation! And I mean everything. *Everything!* Every day of our lives we are living within the confines of our perceptual machinery. Whether we're doing our laundry or flying a fighter jet. It is all a mind construct."

"You know Adam, I've never heard so much crap in my life!" Peter said with some annoyance, "what's got into you that you have to talk like a lunatic!"

Adam chuckled and seemed to be thoroughly enjoying himself.

"I know it sounds weird Peter but it's true. We're submerged in a sea of perception and it sculpts our world. Don't you see? We cannot get away from ourselves."

"You're full of shit!"

"Look, I think you assume I'm trying to undermine human knowledge. I'm not. Science quite obviously works. It's evident all around us. It's got us to to edge of the solar system and beyond. It helps us survive. My issue is with science taking our reality as something that exists independently of us, *in the way it presents itself!* In other words we have no idea of how much we are injecting *ourselves* into that reality. And by ourselves I mean the mechanisms of our mind tools. That's our greatest enemy – the consensual reality. Humans are all built the same; we all conform to a particular design. So we agree with each other as to what 'reality' is. *Big mistake!* Because you see something and seven billion other people see it too, does not mean it has an objective certainty. We're all wearing the same heads Peter, and we all have the same mind tool."

"You keep mentioning this 'Mind Tool'! What are you talking about?"

"The Mind Tool is the architect of our world, Peter, and every human has one. In fact every living creature known to us, has one."

"What do you mean?"

"The mind tool is a living entity's *instrument of cognition*. It utilises a vast multi dimensional vocabulary which literally builds our world. Yes, this place," Adam said sweeping his arm around his head. "And it operates like a lens. It focuses on frequency. Using our five senses it weaves our world into the fabric we call 'reality'. This place. A place that pretends most convincingly to be an external reality!"

"Uh? What are you *on!*"

"The mind tool can only work, or build, if it focuses on a particular energy frequency. And the reality we find ourselves in, *is a result of this frequency*."

"You've truly lost your marbles my old buddy!" Peter scoffed, "I mean 'come on'! You're making this up as you go along!"

"You think so? I'm actually speaking the truth my friend. It's just that old habits die hard. You're only convinced this coastline, that cottage, the cars … are outside you…separate to you, because you've been born into that way of thinking. We all fail to recognise that what we're in fact looking at is a perception."

"A perception validated by other perceptions, Adam. Not just by looking. You may say the wall in front of you is only a visual perception. In which case try and walk through it! And don't blame me for the bruise on your bloody forehead!"

"You're misunderstanding me. I'm not saying that a 'perception' is nebulous, like say a hologram which we can walk through. Not at all. A perception in this situation is an energy structure. But the perception of the structure is not the structure itself. The object 'in itself' or in its unobserved state, is energy of a certain frequency. A frequency which we are highly reactive to as human beings and which our 'mind tool' is obliged to emphasise, in order to keep us alive. The connection with the 'mind tool's' depiction of the 'wall' and what the wall 'actually' is, are probably worlds apart."

"Oh God here we go," Peter said tearing a clump of grass out with his fist.

"Listen, let's imagine we're at Cape Canaveral and been given close access to a Space X rocket. There it is on the launch pad. That towering machine is so obviously an object outside you. *But in actual fact, Peter, it isn't!* You can shout, scream, throw your hands in the air as much as you want but I'm telling you … the object in front of you that purports so convincingly to be separate to you, is actually your own creation, your own language …your own symbol system. Down to the last miniature circuit board, the last tiny screw. It's a mind sculpture!"

"How long must I endure this drivel?"

Adam stared at his friend and smiled. "Look, I'm not trying to undermine the mind-blowing technical achievement that forms the architecture of that machine. The engineering expertise. I'm just pointing out the fact that the rocket is a masterful production of your mind tool. All its components are calibrated by engineers to 'work' together, within the parameters of our energetic landscape. A landscape defined by our own energetic architecture in relation to channel 458, but here's the thing …a landscape which is *presumed to be externally fixed!* It's called the real world! And though it pretends to be, it is *not* external to you – it is a very precise creation of your Mind Tool!"

"You need help mate ….and fast!"

"Peter, the multifaceted mosaic of the mind tool or as I said its 'language', has evolved to be precisely calibrated to the real things or energies out there and the relationship they have with our own energetic architecture. Hence progress and the success of technology! But the real things out there are not the things we're dealing with. All the time we're dealing with our *own system of interpretation* in relation to those real things. And that system of symbols is what we call the world. Put another way, it's the vocabulary of the Mind Tool."

Peter looked at his watch.

"I know you want to go, but please hear me out for a bit, my old friend," Adam implored, "I want to get this off my chest and then we can go down to the beach and search for your precious fossils; it's way too bright to see fossils at the moment."

Peter sighed heavily.

"Look I really appreciate your input ….I know it's annoying. It's just that you're probably the only person I can discuss this with."

"Ok, ok … I'll lie back and study those clouds whilst you divulge your grand theory of the universe," Peter said, lying back with an arm behind his head.

"Thanks ..I appreciate it," Adam said smiling. "Ok, firstly I need you to hold in your head the idea that the universe we inhabit is *not* external to us but is a branch of our perception – a mind language that is reacting to an external reality we can never get to."

"If you say so," Peter said lazily.

"Don't laugh, but what I'm particularly interested in, is the connection of mind and matter."

"Wow … lofty stuff my old buddy."

"Like I said earlier, everything you take to be 'matter' like that car of mine, the ground, that rock, your hand, indeed your whole physical body are just elements of your perception, not made of some externally existing matter, but of your mind language."

"I'm listening."

"What you take to be solid objects are in essence a mind vocabulary. The rocks are its letters, the mountains its words and the solar systems its sentences. The galaxies are its chapters and the universe is its book. And Peter if we step away from this book, I believe something else …something truly profound would come into focus! A library …a cavernous library standing before us containing books without end!"

Peter groaned loudly but Adam continued.

"This is an indisputable truth Peter and yet it is overlooked every day of our lives!"

"What? That we're living in a library?"

"No! That the reality we are living in, is actually a map of our own making! Synthetic. The objects of our world are only symbols Peter. That rock there probably has as much connection to the 'thing' it represents as the word 'apple' does to the thing you eat!"

"Oh for God's sake, please!" Peter said shaking his head. "Are you taking us back to Plato?"

Adam shot him a glance. "I'm not sure where Plato comes into this?" he said raising his brows.

"Didn't Plato say that the world of objects is a poor reflection of the real thing?"

"I have no idea. I didn't really get on with his theory about there being 'perfect forms' which are truly real and the material objects of our world being sort of shadows of these ideal things. I don't think I understood what he was driving at."

"That guy wasn't silly."

"Well, that goes without saying …but it doesn't really help me and to be clear, what I'm saying here has absolutely nothing to do with Plato. I'm not regurgitating his stuff if that's what you're thinking!"

"Okay …just checking."

"Now back to that rock. You are convinced that it exists independently of you, yes?"

"I know it does …. you crazy drug lord! If I chuck it over the cliff and it hits some poor bastard on the head and kills him, would you still say it's made of mind?"

"I'm afraid I would. Because it's made of mind does not mean that the thing it represents is unreal! I just said that the mind tool is calibrated to real things out there which have 'presence' or 'power'. That's why the mind tool recognises them as solid objects. Science subscribes to this, although many scientists have parked the notion in their headlong pursuit of knowledge. Science knows the *subjectivity* of our *receptivity* and yet it acts as though it was an irrelevance. It proceeds in the expectation that study, observation and analysis of these (subjective) phenomena will eventually lead to an overall understanding of everything."

"But they're not subjective. Science studies objective reality, Adam, as agreed by consensus. I understand that some phenomena *pretend* to be objective when they are in fact *subjective*, like an hallucination for instance, but I don't see what you're driving at."

"I'm trying to make you realise that everything we perceive is a subjective phenomenon whether it's an hallucination or a boulder."

"I just don't get what you mean! How can you say that science is studying subjective phenomena? It studies objective phenomena, Adam! I'm not dreaming that rock, or your car! They're actually there!" Peter said in exasperation. He slid his hand in his pocket and brought out his mobile. "Ok, let's see what Google says about phenomena." He tapped the keyboard. "Here we go... *'Anything that can be perceived as an occurrence or fact, by the senses.'"*

"And that perception is a product of our mind tool," Adam said quietly.

Peter ignored him. "Or this one," he said scrolling through the list, *"an object or aspect known through the senses rather than by thought or intuition."* "Or this…in the writings of Kant .. *'a thing as it appears and is interpreted in perception and reflection, as distinguished from its real nature, as a thing in itself."*

"There you go!" Adam exclaimed, "As distinguished from its *real nature …as a thing in itself.* In other words it's saying that an object like my car is a *personal interpretation*. It is not the object as it stands outside of perception!"

"Eh?"

"Like I said, the car we observe is part of the vocabulary system of our Mind Tool. It's not something that stands separate to us. It *is* us!"

"Can you really sit there and say that your car is not an objective reality for God's sake?"

"Yes I can. That car is not an objective reality - it is made of human perception or if you like, awareness. That car does not exist separately to you, in the manner it presents itself. You *could* say that there is no such thing in this world as objective perceptual reality. And you know what? Many top physicists have come to the same conclusion."

"Come on Adam, stop fooling around!" Peter said hotly. "The car is a machine created by us using knowledge …accumulated knowledge… of materials, metals, alloys, plastics, of forces like combustion .. of electro magnetism ..of motion! How can you say that science isn't gaining an ever increasing knowledge of our world and how it works, when right there at the side of the road is evidence ..hard evidence? Just look around you at the incredible achievements of the sciences. We *are* gaining more knowledge of our world Adam …you *know* we are! Look at the extreme engineering feats of recent times. They're putting robotic probes on far away worlds using cutting edge technology. They recently put a lander on Mars, for Christ's sake, that's so damn sophisticated, no single person can fathom it!"

"This is precisely the crux of the predicament we're in!" Adam continued, "you mentioned 'our world'. Science is revealing more and more about the workings of '*our* world, *our* universe'. What you say is absolutely right. We *are* gaining more and more knowledge of our world and how it works - but Peter … *it is not a world that exists independently of us! It is in every sense 'our world' a subjective world, a model created by our minds in relation to a particular frequency.*"

"But you seem to be overlooking the fact that our interaction with this so called 'perceptual world' is being validated all the time by experience. In other words we are learning how to operate in this world with ever increasing knowledge which proves it has external authenticity! It's called empirical science, Adam!"

"You're quite right. It's always validated. And that's its trick! But why wouldn't it be? That's the efficiency of the Mind Tool. You have to realise that the mind tool is calibrated to present reality to us in a very particular way. Why? Because it's primary aim is to facilitate our existence within this reality. And as living entities we have a very precise energy structure or design. The MT's job is to enable us to survive. It therefore presents the 'world' in a way that matches our energy configuration."

"So the mind tool sculpts reality to suit us! What utter garbage! I mean c'mon Adam, how can you possibly suggest this? It makes no sense!"

"It makes complete sense. Rather than sculpting reality it's more a process of emphasising parts and diminishing or ignoring others. To this end the MT presents us with a tailor made reality that maximises our chances of survival within the theatre of this reality. It shows us, as beings with a very precise energy structure, the parameters of survival. But the reality we believe so strongly to be external to us, is not 'The Reality'! In actual fact it's a totally fake place, but beautifully designed to suit our energy configuration. It's heavily biased for our survival, Peter. It's a protector, a dictator and at the same time a deceiver. And I've got a name for it - **Sensorium 458**."

"Oh God please help me!" Peter said putting his hands together and looking skywards. "What in the hell is Sensorium 458? Don't tell me …we're living in a TV!"

"Don't be crass - it's got nothing to do with Tv's or Pc's or radio's or programmes or anything like that. Sensorium

458 is simply the name I've given to the reality we call the Universe. We live in a very specific, 'energetic landscape' if you like …and it looks like this!" He swept his arm around him. "And yes we can manipulate elements of that energy to facilitate our lives. It's the energy that forms our world, and it's the energy that forms our bodies!"

"That energy? What on earth are you talking about? Do you mean the material world?" Peter said.

"In a sense. You need to unhook yourself from the idea of the solid external world we're living in as being a bona fide reality, Peter. It isn't. The 'solid external world' is a symbolic place only. I've named it Sensorium, or more accurately Sensorium 458, to try and emphasise this. It is created by our Mind Tool which in turn is fed by Channel 458. If you like…Sensorium 458 is our 'world'….the Mind Tool is the 'architect' of that world' and Channel 458 supplies the feed."

"Supplies the feed? What are you getting at?"

"The 'feed' is absolutely everything that impinges on our consciousness from the 'outside' *on that frequency.* Remember, there are countless Channels. Our mind tool is focused or tuned to one only or more accurately, to one predominantly. We as living beings, are designed to live out our lives on this channel. And yes… throughout history, and particularly since the industrial revolution, we've gained more and more knowledge of the world, according to 458, and how it operates. But just like you, science believes that 458 is the one and only channel and that the reality we live in, behaves as it does, *regardless of us."*

"Hang on! Science certainly does not think that there is a particular channel that we're tuned to. It has no recognition of bloody channels. That's been cooked up in your mind!"

"Look, I understand you. Science generally thinks there's one reality and one reality only. But I disagree. I've only called our 'reality' a channel, in order to suggest there are others …on different frequencies. It's just a device. I'm convinced we live in a frequency, Peter. And this frequency, with the use of our Mind Tool, creates this… the so called, Material World." He swept his arm around him. "I suppose you could say that our mind tools enable this energy frequency to masquerade as matter."

"Look, that's nothing new. Physics knows that matter and energy are one and the same."

"Oh? So we've suddenly changed our tune?"

"Not at all."

"Sounds like it!"

"No, I was just repeating something that was drummed into us at school. That's all. Matter and energy are one and the same."

"Easy to say, but what exactly does that mean? Are you now agreeing with me that the everyday world is a mind construct with no true validity?"

Peter looked towards the ocean. "No I'm bloody not! I'm…I'm really not sure what it means in terms of our experience of the world. I just can't stomach your idea that the world is made of mind and isn't something objectively durable … something existing whether we are here or not! That whole idea is bloody preposterous!"

"Do you think I don't agree with you? Of course it's preposterous. It's bloody absurd. But we have to be honest. The world we live in does not exist independently of us!

Even stranger is the idea that if our minds created this place using a particular frequency, what else could they create when locked on to a different channel?"

"Hang on! I cannot take this channel business. That really is out in sci-fi world!"

"Really? You've already admitted that this place is basically made of mind, so what's the big deal?"

"I did not!"

"You bloody did. You said that matter and energy are one and the same so what exactly is this place?"

"You know what … you're bloody exhausting Adam! Stop twisting my words. I absolutely know this world we live in, exists, but hey, why don't you just talk and I'll listen."

"Right my old friend, this is my take."

Peter snorted.

"Our world is a mind construct, fed by a frequency, a particular channel … and from birth we are locked onto it. Let's call it an energetic landscape. It pretends it's the one and only but this simply isn't true. Other landscapes are right before us, but because their frequency is slightly different, we just don't see them!"

Peter groaned. "What on earth is an energetic landscape?"

"It's not *on* earth Peter, earth is part of it; and it stretches as far as you care to direct your attention – in other words to the edges of the universe and beyond! It's really a virtual map, created by the mind tool that defines the viable environmental parameters for any particular living entity, thereby presenting it with possibilities of action."

"Sooo …the energetic landscape is defined by the … Mind Tool?"

"Absolutely. It's derived from the relationship between the energetic structure of a living being, whether it's a beetle or a human, and its surroundings. The mind tool uses this relationship to generate a virtual world for its owner. We're sitting in ours right now. The EL is the same for all living creatures known to us because they all subscribe to the same fundamental energetic composition and are tuned to the same channel. But their virtual world will be very different for survival purposes. In other words it will have strong biases according to their lifestyles."

"So when you say all creatures known to us live in the same energetic landscape but their virtual world is different, what do you mean?"

"I mean that all creatures known to us live within the walls of the 'material world' as we understand it. In other words if you put a mosquito in a jar and screw the lid on, the mosquito is trapped."

"And?"

"And this is true for all creatures we know. In other words they are subject to the laws of sensorium 458. But the virtual world the mosquito inhabits is vastly different to our own due to its lifestyle. Its mind tool will have created biases for its survival and therefore a symbol system, tailor made to its survival. Does this make sense?"

"I suppose so. You're saying the objective energetic landscape is the same for us all, but the virtual worlds of the different creatures on this planet are different, to favour their survival?"

"Precisely! A virus, an ant and a mouse will all be trapped within the confines of our hypothetical jar, but their 'mind map' of that jar will be very, very different."

"But when does a creature inhabit a different energetic landscape?"

"Bingo! When that creature's mind tool is focused on a different channel! And that's when the energetic configuration of that creature is different to our own and indeed different to every creature known to us at the present time."

"I'm trying here …I really am. When you say a creature that is unknown to us, I'm guessing you're referring to …to paranormal entities?"

"Of course. Entities that science is quick to dismiss as fantasies despite the plethora of information to the contrary."

"Ok."

"So you go to put this 'new' type of creature in a jar and guess what? It doesn't even recognise the jar and it certainly wouldn't be held within its confines because it's not part of that creature's energetic landscape. The barriers set up by the jar are irrelevant to it. In other words the physical constraints of its existence are entirely different to those that exist within the walls of Sensorium 458!"

Peter fell back on the rug, moaning, with his hands over his ears. "I can't take this …it's just too much!"

Adam laughed. "Sorry buddy. You have to realise there are many other worlds which exist right here, right now, all around us, built from differing energy frequencies and creating their own landscapes. And they're as hard and convincing as this one. Every world has a frequency and do you know what stops you recognising this? The *illusion of solidity!* Until you take on board the fact that 'solidity' is a feature of your perception or mind tool…. as opposed to the 'environment', your understanding will always be blocked. How can another world exist here when you've got this!" Adam said, slapping his hand hard down on the earth with a thud.

"Exactly!" Peter said.

"It's impossible right?"

"Right!" Peter affirmed.

"Well sorry … actually NO!" Adam said fiercely, "Everything is energy and our minds are designed to form worlds from energy. Take a look at our world, Sensorium 458. It's unbelievable that our minds can create this magnificent place….this all consuming, complete world. And our Arena is primarily everything that pretends to be matter and indeed the space between that matter. Your hand, your body, this cliff top, the sea, the Hubble telescope, the wasp on your cup, the moon, the sun and all the galaxies of the universe are ALL ….SENSORIUM 458!"

"You're really quite insane Adam aren't you? There's a mad worm in your mind! I'm sort of worried you believe what you're saying," Peter said, propping his head up on his arm.

"If seeing through a trick is insanity, then yes I am insane!" Adam said light heartedly, "but is what I'm saying really *so* insane? I mean what could be more insane than accepting as possible, the instantaneous collapse of a gigantic star, containing trillions upon trillions of tons of iron, into a tiny point with no volume? You accept it because a bunch of smart guys called astrophysicists made some calculations, did some observations and told you its true. But Peter, on scales of madness that's got to rank with the best of them! That's as impossible as what I've just told you."

"Are you saying you don't believe they exist?" Peter asked

"Of course I'm not! I'm using them as an example of something truly insane that flagrantly does exist …in the arena of Sensorium 458 of course!" Adam said laughing. "Can you imagine your bathroom shrinking to the size of a pinhead….then your whole house …then your entire street …and city ….and continent ….. and entire planet? And that's just a speck of the actual amount of matter that gets swallowed. That in itself tells us that matter is indeed something very different to the 'stuff' that presents itself to our minds every day of our lives."

"It's down to massive gravitational forces Adam, that's all. Electrons are forced out of their orbits and have to amalgamate with their nuclei," Peter said matter of factly, "and then even more extreme events occur at sub atomic level to allow the formation of black holes."

"Oh well that explains everything," Adam said, "except it doesn't quite does it. The equations go a bit barmy in singularities … they break down and the physicists are left scratching their heads. In a nutshell Peter, Einstein's theory of relativity goes a bit goosy. They don't know what the hell is going on, which is why you shouldn't scoff at the idea that we're living in a particular reality I just happen to have named Sensorium 458."

"How does that explain anything better?" Peter said perplexed..

"I'm not saying it will explain anything any better ….I'm just asking you to be open minded. We are living in a particular slice of a much grander reality; we are under the tyranny of our frequency. It pretends it's the one and only, but it's a deceiver. We are trying to understand this higher reality through the meagre resources offered by channel 458."

"Ok ..ok," Peter said drawing himself up and facing Adam, "to get this straight you're saying that we only pick up or perceive energy at a certain frequency and that our world is predominantly built of it? Is this right?"

"Not quite. Our world is built of mind - a super complex arena of symbols. The builder of this arena is our mind tool which, if you like, creates a symbolic reality through its interaction with the raw code…Channel 458. Our world, which I've referred to as Sensorium 458, is not a 'place' and never has been. It represents a presumed reality if you like. Sort of like a robot banging into a rock and it's 'mind' lighting up with red lights. The lights are not the rock. In the same way, Sensorium is not the 'World'; it's a symbolic world. Channel 458 delivers the information of that world which our mind tool gets to work on and creates the architecture of our 'reality'…Sensorium 458. And this is where we are right now, sitting on the grass on this clifftop."

"Adam, you're bonkers mate. You're seriously sitting there, telling me that this world doesn't really exist?"

"I'm sorry, but no, and you agreed with me! We're sitting in a mind construct right now. But a bloody brilliant one eh? It's got you fooled! Do you realise that the whole of human history has been conducted within this arena? No one has ever been in the 'real' world Peter. At least no one normal. Just … shall we say, psychologically advanced humans."

"I don't find this particularly funny."

"Look ….all I'm saying is that our reality is the interface between our 'mind tool' and this order of energy. And when I say reality, I mean the solid trusty world we all stand in – it's form if you like. Because it's the physical presence of our world…. the mountains, the seas, the cities, which gives our world its most powerful sense of realness."

"So this particular form or frequency of energy which we have labelled matter, creates the world we inhabit?" Peter said.

"No. The Mind Tool creates our reality by focusing on this frequency … Channel 458. We react to it and give it the qualities we define as matter along with the myriad other components that form our world. To put it another way, the interaction of this energy and our mind tool results in the perception we call the world. And we, as 'solid' objects in Sensorium, are a part of this vocabulary."

"Go on."

"And this perceptual reality we live in, is fine tuned to facilitate the existence of the beings we are. But as I just said, the beings that we perceive ourselves to be and the reality we perceive ourselves to be living in, are only a model of a reality we cannot reach. It is not real! Together they create the world we live in. Imagine you were born blind and you're being looked after by a pair of dictatorial aunts who shield you from life by locking you up in a huge house. You wouldn't have a clue! It would all seem perfectly normal. This is what the MT and channel 458 are like - a pair of dictators steering you every step of the way.

And I'm using the term frequency to denote a particular quality, presently outside the reach of science".

"Perhaps because you're not a scientist?"

"That's probably to my advantage, Peter. My thinking isn't in a straight jacket. The world that much of science is understanding is an illusory reality which has more to do with the perceiver than the perceived. Science is in thrall to the vocabulary or symbol system of our Mind Tool …and it won't have anyone tell it otherwise."

"Don't dismiss it Adam. One day this so called illusory knowledge may save your arse in a hospital or pluck you from the deck of a sinking ship in an illusory aircraft!"

"Look, I can't argue with your point, because you're absolutely right. We are slaves to 458 and as such we live and die under its dictums. Blood, muscle, bone and pain are the bricks which form the reality of this world ….but you have to understand… *they're made of mind!* We can only fully appreciate the nature of this place when we can escape its clutches."

"By dying?"

"By dying out of this world"

"Well what's the difference?"

"Dying out of this world could mean a shift of perception. A discarding of the energy we have accumulated all our lives and which forms our 'solidity'… our earthly vehicle if you like, and refocusing."

"Refocusing with what?"

"The mind which stands separate to 458 and is therefore incomprehensible to us in this state."

"Our spirit then?"

"You want to label everything Peter - spirits and souls are a contrivance of 458. They stand as nebulous entities against the harsh solidity of our world, when I've been trying to impress on you that our world is far from solid."

"Your logic is cockeyed!"

Adam sat silently for a few moments gazing out to sea. Then he spoke quietly. "And yet you're happy to believe that the whole universe was at one time compressed to a point with no volume… something way way smaller than an atom. Now who's logic is cockeyed?!"

"Oh God Adam you're annoying!" Peter said throwing the stone over the cliff.

"Well I'm sorry … it's just a conversation I've been longing to have and I knew you would enjoy it!"

"Enjoy it, my arse! Anyway what is this Order of energy? What the hell do you mean. There are countless types of energy in our world as it is, so where does this 'order' fit in? Have you suddenly become a theoretical physicist, a quantum mechanic and an astronomer all rolled into one for God's sake?" Peter said incredulously, "and how does it allow for the existence of other worlds within our own?"

"As I said, this particular energy frequency gives our world its solidity or mass through the mediation of the mind tool. If you like it forms the 'realness' of our world."

"Hang on, you just said our minds created the matter of our world!" Peter said.

"That's what I just said! By locking onto this particular energy frequency. That stone is made of it, so are you and me. But there are other entirely foreign orders of energy all around us which remain hidden. But if I could alter my tuning and pick up a different frequency, my experience of that stone would be entirely different. In fact it would no longer have the properties we attribute to it and it may simply vanish. But out of nowhere an entirely foreign world would appear and at the same time I would no longer be the being I was – I would have changed."

"My God Adam! Where is all this insanity coming from? Worlds appearing from nowhere…out of thin air? We're both turning into a pair of nutcases!" Peter said with a snort, "it's all a bit much this."

"Peter, all the forms of matter that are specified by the periodic table and which give our world its solidity, come under one Order of frequency. And we are tuned to it. And it is 'we' who inject the quality we call solidity into that energy….and create the hard world we live in. Solidity is not intrinsic to that energy Peter ….that comes from us. If you like it's a mind trick to facilitate our survivability."

"Oh please! How come if we trip over a rock we bruise our bloody leg?"

"You're sounding like Samuel Johnson trying to refute Bishop Berkeley …with a crass piece of logic!"

"I am?"

"Look the answer is simply because the energy called a rock, impacted on our own energy and damaged it. The sensation of hardness is merely a mind symbol to allow us to live in this landscape."

"Yeah right!"

"And so it follows that were we to tune ourselves to one of the other Orders of energy which exist all around us, then we would find ourselves in equally hard, but entirely alien worlds!"

"God help me!"

"And probably with very different survival parameters."

"Look … to get back to the point, you seem to be implying that it's all a matter of perception. That the stone has no real solidity in itself, only in its perceived state"

"Bang on the money!"

"But Adam ..c'mon, we both know that the stone exists independently of us don't we!"

Adam sighed. "The energy that the 'stone' represents exists independently of us, sure, but the stone we perceive is a mind construct."

"Look, let's be clear. The stone, the rocks ..the beach are here whether we are here or not!"

"How do you know that?"

"Because it's bloody obvious! Are you out of your skull Adam?"

"How do you know these things …what makes you so damn sure?" Adam said staring at him.

"Is there some way of knowing these rocks are here if we're not here to clock them?"

"By setting up cameras … weighing machines ..for fuck's sake. Adam are we really having this conversation?" Peter said in exasperation.

"As you know Peter, cameras and weighing machines are only extensions of our perception……enhancements of our senses. These things prove nothing!"

Peter groaned.

"Do you understand now how impossible it is to escape ourselves?" Adam continued "You say you know the beach has an objective existence …because other people can verify it …it's always here when you arrive …it's on a map …it's on satellite imagery, but I think you're beginning to realize Peter that all these recordings, measurements, witness verifications bla bla … are all perceptions, and therefore are under the jurisdiction of channel 458. You can't creep up on the beach to see whether it really does exist separate to you …you would first have to climb out of yourself. Don't you get it? We're all infected by this false concept of external or objective fixation."

"God help me"

"Look Peter, imagine a man standing in a cathedral. But this man is rather unusual. He can only see the glass of that building - the stone is completely invisible to him. After a period of time studying these glass panes he begins to realize that they shouldn't really be there …cannot be there, without something to support them. He realizes something is just plain wrong and makes an outlandish suggestion to his friends that there's a lot in that place they can't see. Now that guy may swear blind that the glass cathedral exists whether he's there to see it or not, and his friends, who are all the same as him, absolutely agree. They can verify his observations. They can leave the cathedral and come back later to see it standing there again. But who's right? Is the glass cathedral real or is it false?"

"Spare me Adam ..please!"

"One could say that it does indeed exist, but only if one inhabits the cognition of this man and his friends. From a fuller perspective however, the glass cathedral is an utter fantasy - without the stone that forms by far the greater component of the architecture, it presents an entirely false picture! And there's additional deception here. The glass cathedral is itself only a symbolic structure ...a mind creation, so not only is our man only seeing a part of the whole, he's also transforming that part into his own perceptual language!"

"For Christ's sake Adam ... say ... say an asteroid were to strike the earth like we've seen evidence of them striking all over the solar system, then all hell would break loose yes? In other words it's not just a perceived solidity it is an actual solidity."

"Peter, everything is perception. You say the asteroid must have real, not just perceived solidity for it to do the damage. As I said earlier, like the wall, it has power and presence – a quality we call real solidity. But *real solidity* is at its core a quality derived from perception, and our perception is under the jurisdiction of the mind tool as it fixates on channel 458. And in the world of 458 that particular energy frequency is highly reactive to the beings or structures that we are. So yes... *our world* would be decimated. But 'our world' is not *the real world*".

"Bollocks!"

"Your problem, Peter, is in understanding the nature of solidity or hardness. You've convinced yourself that they're objective qualities. I'm saying they are not, but that is not to say that they are not real and deadly and utterly capable of dictating the course of our lives within the Arena of 458."

"But I still don't understand" Peter continued, "if they're not real, then how do they do any damage?"

"They are real in a sense. It's your understanding of real, which is the problem. They are representatives of very real things but they are not those things themselves! These asteroids are made of the same energy as you and me and within the walls of 458 the rules by which we interact with that energy are very precise. In other words if you walk over there and jump off the cliff you'll interact with that energy in a rather detrimental way! You would be transgressing one of the survival rules of our reality. The MT would give it a more flowery description …like a load of guts splattered over the rocks!"

"You're a bloody nut case, Adam!"

"You have never seen the bona fide world Peter and nor has any other normal 'human being'. From the moment we're born to the moment we die, we live in the vocabulary of the Mind Tool! But all the time we make the mistake of believing we are learning more and more about an objectively existing world. But as George Berkeley said, 'all we perceive are our perceptions.' That's what Kant meant by 'the thing itself.' The noumenon. 'Something' that is just itself… without being perceived."

"I don't agree," Peter snapped, "we *are* learning more about an objectively existing world, but not necessarily through an act of perception. A lot of physics has absolutely nothing to do with perception. It's deduced from mathematics. How do you think they worked out how the sun burns….by putting a slice of it on a slide under a microscope!"

"Ha ha," Adam said, "that's funny. Look Peter, you've actually made a very good point there.

Mathematics is like a torch shining in the dark. It's a true seeker of truth and it finds the truth. The problem is that the owners of this torch cannot comprehend what the torchlight is revealing. They're so steeped in the edicts of the mind tool that at certain junctures, the mathematics makes no sense."

"So you say, Adam!"

"Look, I am not disputing that the laws of physics are often built on theory, intuition and mathematics i.e. not on direct observation. But right at the core of these abstract calculations is the platter laid out by our friend, the mind tool. Like the rotation of the galaxies. That forms the bedrock of the thinking and then the theory follows. And then, how are these theories ultimately proven to be accurate or inaccurate? Through observation and measurement of phenomena! And guess what …channel 458 and our mind tools have complete control over phenomena i.e. how 'stuff' presents itself to our consciousness, and that is an inescapable truth".

"Stuff?"

"Okay, reality then. You're confusing yourself with words. Stop thinking of stuff as matter. Einstein and others, mathematically calculated the existence of black holes through mental cogitation. His common sense (his grounding in 458) told him that their actual existence was unlikely if not impossible. But guess what.. channel 458 revealed that the monsters actually existed! Mathematics had poked its finger through the skin of our world and made an impossible discovery! Something was just plain wrong!"

"But the particular energies you talk about that make up these foreign worlds …how in the hell do they escape us"? Peter said indignantly.

"In the same way that dark matter does. Our senses can't pick it up or don't pick it up – it's an unknown order or a whole bunch of unknown orders with peculiarities which make them invisible to us - we can only infer their presence," Adam said, throwing a clod of earth towards the cliff edge. "That's the energy we are designed to perceive and which forms our world and our universe. It's hard and convincing isn't it. Dark matter may be as hard as iron to a being on a different channel…or a whole string of different densities of an alien energy. The strange behaviour of the galaxies … you know…the lack of observable mass is probably an example of our 'slice of reality' revealing that it is only a slice and not a complete system."

"Go on."

"The inconsistency might be caused by interference … the presence of an energy frequency very close to our own but nonetheless invisible to us. Such as you might find on neighbouring channels. And I don't mean neighbouring in a spatial sense Peter. You absolutely must understand that the universe is 458's presentation to our minds …it is not a place. It's made of our perception. Mars isn't a place, nor is Venus, although they appear to be …nor is the black empty vault in which they spin. They're perceptual realities on our channel, in the same way that the glass cathedral is a perceptual reality to our peculiar man."

Peter burst out laughing. "Well how the hell can we land a spacecraft on them?"

"You may as well say the ground is real because you can place your foot on it! It's energy, Peter, energy which we are tuned to and which forms part of our energetic landscape.

The mind tools' presentation of reality is built on a relationship or reaction between us and the energetic landscape we inhabit. It presents reality as something separate to us, but it is not. It presents the objective world as a place, but in truth that place is the *physical manifestation of an interaction*; a very precise interaction between the energetic landscape we are tuned to and our own energetic configuration."

"You're doing my head in, Buddy."

"Now you see how the mind tool is our navigator. Its job is to facilitate our passage through the sea of reality by producing a make believe world that enhances and diminishes particular features for the benefit of good navigation. Everything we know about Mars, Venus, Jupiter could alter in a blink if we could change channels. Dark matter is probably alien energy - a whole spectrum of different energies which reveal themselves on different channels. I think we are looking at a tiny slice of a mega reality …at God's eyelash if you like."

"Are you going religious on me?" Peter said mockingly.

"I don't think so." Adam replied. "The problem is that we're looking at this from different angles. You see the universe with its attendant galaxies as something objectively existing – trillions upon trillions of balls spinning around a limitless black void. I see it as a perceptual construct of the human mind …as Sensorium 458's rendition of an invisible reality. There is so much missing. How do we know that there isn't a whole mass of unseen energy out there which is having absolutely no effect on our universe and therefore cannot even be inferred? Matter is a mind invention Peter as is our universe …it's an artificial sensation, don't ever forget that."

"I've just never heard such a heap of unmitigated crap, Adam," Peter said drawing his knees up to his chest and looking out to the horizon. "Sorry, but I can't stomach this!"

"And yet Peter," Adam said smiling, "it's funny how you're prepared to accept some madness but not others."

"Like what?"

"Like black holes ... singularities ! Why is what I'm telling you any more insane than the concept that all the matter in the universe starting from an infinitesimally small point? Is it by any chance to do with a school boy reverence for men in white coats and blackboards crammed with equations? Think about it Peter …a single tiny point!! How insanely mad is that? Nothing I can tell you even comes close to such an absurdity and yet with a shrug of your shoulders you gulp it down!! It only sounds possible because of our utter incapacity to understand…..to really grasp what it actually means."

"Ok .Ok. But how can these worlds all exist so independently … be so insulated from each other? It just doesn't make sense."

"You have to realize that the walls between the worlds are formidable."

"You mean the channels?"

"Yes. They're built of energy disparities which science will one day come to recognize," Adam said matter of factly.

"Adam, I'm really sorry but you're not a scientist – you don't have the right to make statements like that if you can't back them up with some sort of factual evidence!" Peter said with a serious expression.

"I can to you Peter. I'm not preaching this to the Jet Propulsion Laboratory."

"Go on, I dare you!" Peter said smiling broadly. "They'd buckle up laughing!"

Adam chuckled. "You're spot on there. Those people are the high priests of 458. Precision minds working within the vocabulary of the mind tool and pushing the technical boundaries of our world. Unless you have a string of letters after your name your thought processes are invalid."

"Look … get back to the point … this business of 'other worlds' being around us. There has to be proof. That's the story of scientific discovery"

"And that's why science in a way remains blindfolded – it demands evidence in a format that keeps it trapped within the theatre of sensorium 458."

"The scientific method is massively circumspect Adam – hence its success. Often the phenomena that science observes, run contradictory to its expectations …like the expansion of the universe. Astronomers expected it to be slowing down but in fact it was speeding up!"

"But science still holds certain falsehoods dear – like external durability or fixation. Tell an astronomer that Venus is a mind construct and he'll tell you you're insane. *But that astronomer has absolutely no proof that Venus exists separate to him!* That is taken as a given. But the plain fact is that no one has ever experienced Venus outside the theatre of Sensorium 458. Or to put it another way, outside of human perception. As I've said, our eyes are not windows onto a real world!"

"Hang on, he *would* be able to say that Venus exists independently of his observation, because Venus affects other planets in the solar system and it would therefore be a matter of deduction!"

"You're simply not getting it, Peter. Of course you can deduce the presence of another planet by the reactions of those around it ….*because they all belong in the same show!"*

"C'mon Adam, you only have to sample the rocks and strata of those places to know that they existed way way before we humans came on the scene. So how can they only exist in the minds of an observer when they blatantly existed way before the 'observer' was able to observe?" Adam said, raising his eyebrows.

"Those rocks from long ago are all being observed by someone through the lens of Channel 458. A human mind. A mind geared to that particular channel. They are all phenomena on that channel. Mars exists in that reality …and yes of course it has a history. But it is created by our mind tool, so of course all phenomena will dovetail together in a logical way! Mars is real …totally real, in sensorium 458, because that's what the MT demands, but on different channels it will alter dramatically or simply cease to exist. That's what fools us into thinking we live in a complete world – everything …or almost everything, ties in together."

"You're losing it mate. You're speaking gibberish and yet you're convinced it makes sense! You speak of this 458 place as being a certain category. What the fuck are you talking about!? Where are these so called other categories?" Peter said throwing a stone towards the cliffs.

"Look Peter, there's a little subtlety in here that you're not getting. It's obvious but at the same time very elusive.

We live our lives inhabiting the vocabulary or language of our mind tools whilst believing that we're living in a real world. Think about it enough and just like those trick pictures which can be seen from different perspectives, your mind does a back flip and suddenly you're seeing a different picture. Just remember that even the hardest thing you can encounter is built of mind and then it all falls into place."

"Ok, I will think about it and let's hope for your sake I have a sudden revelation! If I don't you're off to the loony bin!"

Adam laughed and slapped the rug. "Look Pete, would you admit there's a lot of very strange stuff in the world as it is?"

"What do you mean?"

"It seems to me that you're closing your mind to my ideas when there are some mighty odd things in the 'sensible' world you subscribe to."

"Like what?"

"Like the fact that the quantum world has the very weird quality of changing according to the mindset of the observer. It cannot separate the *observed* from the *observer*. Einstein once commented that the moon does not exist unless one is observing it!"

"Rubbish!"

"Which? That he said it or the idea itself?"

"Both. He neither said it nor is the idea valid!"

"Actually he did, my old friend, look it up. But let's not get excited about it. There are so many things. Like the words of a physicist called Pascual Jordan. I don't remember it so let's see …"

Adam brought out his phone and flicked through various windows whilst Peter swatted an irritating wasp from his plate.

"Here it is …I kept it just for you!"

"I bet you did!"

"No seriously, it's quite profound. Ready?"

"Shoot," Peter said, gazing out to sea.

"This guy worked with a quantum guru called Niels Bohr in the '20s. Here's what he said. *'observations not only disturb what has to be measured, they produce it. We compel a quantum particle to assume a definite position, so in other words, we ourselves produce the results of the measurements*! How crazy is that?"

"Crazy indeed," Peter muttered, "I don't think I feel safe where I'm sitting any more!"

Adam laughed loudly. "And this stuff is part of the 'sensible' real world of science, Pete. Proper trained physicists …the type you really believe in. Not marijuana puffing new age hippies whose brains are cooked! These guys probably started out thinking there's a good old solid world out there which we'll get to the bottom of and then … they realised, it wasn't making any sense. There's something very peculiar about this whole objective and subjective business."

"It certainly seems that way but I guess you've picked out extreme examples. Let's face it, the bulk of the time these experiments produce results that work within the parameters of the world as we know it."

"Sure, I agree there. But push it too far and weird stuff starts happening. Look, here's another. The double slit experiment where photons of light appear to go through two different apertures at the same time…."

"That's old hat … heard that one!"

"Or electrons that seem to be in different places at the same time. You know …Quantum Superposition I think they call it."

"Heard that as well!"

"Ok. I get the message. But even if you have heard about them, they're still bloody odd! And again, the process of observation seems to affect the results."

"Ok."

"And what about the phenomena of objects disappearing and then reappearing sometimes months later?"

"That's people being absent minded or taking drugs!"

"You're such a devotee of Sherlock Holmes, Pete! Anything that doesn't fit in with your cosy world view isn't valid. You should be aware, there are countless records of objects disappearing and then reappearing in bizarre places. Like keys missing for months suddenly found lying on the owner's pillow."

"There are lots of crazy people out there Adam, that's the issue!"

"A lot of these people are far from crazy my old friend. They're just reporting extremely odd occurrences. But science doesn't approve because it thinks it knows the parameters of reality. What's really happening here is that science is trying to lock a ghost in a bottle. And when the ghost shows that it doesn't like being imprisoned, science literally turns a blind eye."

"Don't you prefer a measured, sensible approach, to a gullible one?"

"There's nothing sensible about denying a real event, Pete, even if it does seem outlandish."

"I prefer caution, I suppose."

Adam stretched and took a swig of water from a bottle. "Look, the things I talk about are places where our mind tool reaches its boundary. It can't deal with the information. This is where the term 'paranormal' springs from. Our ideas of normality are challenged. And yet it all makes sense, Pete, when you realise that the outside 'world' is actually a confection of the mind tool!"

"You think?"

"Yes I do! As I said, human beings are receivers on channel 458 as are all life forms known to us. You could imagine Channel 442 and 463 as neighbouring frequencies and although completely invisible to us, these channels form worlds as complete and insulated as our own. But interference does sometimes happen between close channels"

"My crazy God!" Peter said staring at the sky. Adam ignored him.

"That's where strange unexplained phenomena happen ie a person has an extreme emotional event, has a fit or experiences a severe fright. They can then briefly hook into another frequency and experience a foreign world. Ghosts may belong in this category and probably some UFO sightings. For some unknown reason the walls between the worlds can waver briefly and allow non sensical perceptions to filter through. Ironically this is precisely where science steps in, holds up its hand and says, 'Sorry that's a subjective experience …it doesn't count'".

"Pleeeease!" Peter muttered, "aren't we a bit beyond ghosts and ghouls!"

"Oh Pete ... that was so '458'!" Adam said, bursting out laughing, "complacent, bull headed, and dismissive!"

"Well why not? Look how science has improved the human lot," Peter said indignantly.

"Hey … relax and hear me out. Just take our reality as one of many. We're in a unique arena. We'll travel out into space not realising that we're travelling within our frequency. Everything will be '458'. The civilisations we encounter, the planets on which they live."

"Yeah ..I need a holiday …does channel 458 do a brochure?"

"Ha ha! But there will always be strange inconsistencies where the mathematical brilliance of our physicists takes them to places they cannot comprehend. These are the edges of our reality. Mathematics will want to pull them out of their channel and into another, like a fishing weight drags a hook into the depths. But the hook (their sanity) will simply not grasp what it's witnessing."

"Sure ..all hooked up and fucked," Peter muttered.

"The only way they can make sense of it is to work on themselves, on their psyche - and this will never happen because science prides itself on studying the external world in an impartial objective way and anything that smacks of subjectivity is anathema to it."

"Bloody good thing too!"

"But is it? When we've just said that the observation of 'things' has a crucial part to play in the whole process. The true irony here is that the objective world is in fact as subjective as a headache - both are made of the same thing – mind."

"As you keep telling me!!"

"Peter, the only way to gain a fuller understanding of reality outside the walls of 458 is through tampering with our 'advanced settings' if you will. And there are people who are engaged in this as we speak. You can call it religious but in truth it's a very pragmatic system of instruction, designed to destabilise our fixation on this one channel. It bestows on the practitioner very unusual abilities, let's leave it at that."

"So it's a form of hypnotism?"

"Not at all! It's psychological teaching. Look, I'm not a practitioner of it so I can't tell you how it works. I called it tampering with our 'advanced settings' to equate it with pc's …something we're all familiar with. Just accept that it involves mind techniques over a long period of time with the primary aim of destabilising the focus of the mind tool on 458."

"Sounds dark."

"Well it's not. There's nothing sinister about it. It's not for malign purposes that people embark on this learning, it's for freedom. Personally I wouldn't have the guts to do it. There are things out there which are truly terrifying."

"What do you mean?"

"There are things around us which are best left alone behind the barriers set up by our Mind Tools. Disturb its focus and you're exposing yourself. You're cracking the cocoon of your world, if you like, and allowing access to yourself. That's where you have to be ultra strong in your mind - and where the years of teaching come in. It wouldn't be for me. I truly believe that we are surrounded by other worlds and I also believe that we brush with foreign beings or entities all the time. I've called them USB's."

"USB's? What the hell are they?" Peter said, as his eyes tracked a noisy lark.

"Unidentified Sentient Beings."

"Ghosts?"

"No, life forms as alive as you and me but on a different frequency and living in an entirely foreign world. They're as solid as us (within the framework of their own channel) but only to a receiver tuned to that frequency. Don't forget, solidity is a quality created by a receiver (a being like you or me) in order to exist in its particular world - it is not an objective quality. You are not solid and nor am I. We give the quality of solidity to each other through the use of our mind tool."

Peter groaned.

"There are probably many categories of so called 'Paranormal entities,'" Adam continued, "You have the house 'ghost' which flickers into our reality and then vanishes. Then you have the scary 'poltergeist' which actually interacts with the energy of '458' or in other words 'physical objects'. The ghost is a USB which is interpreted by our mind tool into something familiar like a person or animal. They seldom last because they don't exist in '458' – they are a result of our focus temporarily drifting into their reality or frequency. After all they usually only last a few seconds and then dissolve as our minds regain their correct tuning. I sort of imagine that the likes of poltergeists are caused by the opposite - incoming interference from another frequency into our reality. I believe this incoming interference is sometimes very purposeful."

"Are you trying to give me the creeps?"

"Not at all," Adam said, "just discussing real things that most people push under the carpet as unproven hocus pocus! Look, poltergeists could well be foreign beings from other channels which have entered our reality of their own volition. I say this because they actually *interact* with the energy of our reality in a physical way... throwing plates, opening doors, drawers and so on whereas ghosts do not. But these are only the ones that we observe. Many others are here but are way below the radar."

"Jesus."

"Imagine you're a radio always tuned to channel 4 . This is your reality. Imagine there's a foreign being, tuned to channel 2. This is its reality. Generally all works fine with you in your world and they in their world, oblivious to each others existence. But then suddenly the receiver wobbles momentarily and channel 4 radio starts picking up channel 2. It's confused ...the programme is utterly unfamiliar. This is what I mean by a person in a highly agitated state temporarily shifting focus and witnessing extraordinary things!"

"I'm lost for words my old friend."

"But this analogy can only ring true if you accept that our reality is anything but solid! And that's nigh on impossible right?!"

"You said it, Adam."

"Have you never heard how weird phenomena sometimes happen around babies or young children?"

"Possibly"

"That's because their focus on channel 458 is still loose and they unwittingly access other channels. They can then cause odd things to happen because they are not contained within the walls of 458 like you and me. Their shifted focus causes disturbances."

"Look Adam, you're off again!" Peter said sitting up suddenly and wrapping his arms around his knees, "you're talking as though this whole damn business is a truth when the fact of the matter is that you don't have a shred of evidence to support it!"

"A shred of evidence to support it? Don't be ridiculous! There's loads of evidence out there but 'science' has been quick to dismiss it as 'new age hippy bla bla' because it has particular 'rules' as to what is real and what is false. And those rules, as I just said, keep it blindfolded. Back in the early nineteenth century if someone said that rocks can fall from the sky, they were laughed at! Anyone with any sense knew that rocks don't fall from the sky and anyone who said they did, was an idiot! But when three hundred people saw rocks fall from the sky in Normandy in 1803 and were able to pick them up for scientific analysis …well conventional science had to eat its words. Wind forwards two hundred years and guess what? Everyone knows that rocks do indeed fall from the sky!"

"Well, you could say that that is a prime instance of the scientific method correcting itself! It presumed something, and then evidence appeared that contradicted it. So it changed in response to the evidence," Peter said staring at Adam.

"Well, it had no choice! And it illustrates sciences bull headed presumptuousness. But there are numerous examples of 'things' which have been witnessed by perfectly sane people which are not so easily proven. Like UFO sightings, ghost sightings, etcetera."

"Okay, so these different worlds you talk about on different frequencies - where the hell are they and why has science never detected them? How could worlds like ours exist here and remain hidden to an advanced civilisation like ours. I just can't believe that's possible!"

"Believe me it is," Adam continued, "the worlds are insulated from each other by their unique frequency - quite like the millions of mobile calls which are happening at once all around us. The air is thick with them and yet they seldom interfere with each other. Your problem with accepting this is because of your utter belief in an objectively existing solid world. Your utter belief that you know what can and cannot happen. When confronted by something out of this world we scrabble to explain it in our own terms. We think we're being sensible, logical, nobody's fools …when in fact we're just being small minded."

"We have to be careful what we believe, Adam!"

"You know even Einstein himself didn't think black holes would actually exist even though his equations told him they did. Early believers in their existence were actually laughed at by their peers. Do you blame them? That's what I mean about mathematics taking us beyond the edges of our reality - it pierces the skin of our false arena and underlines the incompleteness of channel 458 as a total world."

"Hang on," Peter said, fixing Adam with a beady look, "Black holes have been in the scientific domain for years. They're exotic I agree but not so outlandish as to suggest …well, the existence of hidden worlds. The physics is well understood, it's just a question of time before they disentangle the precise forces underlying them."

"You only speak like that Peter because 'black holes' have infiltrated our language and become sitting room small talk. But if you think about the nature of those beasts for any length of time and really imagine something thousands of times bigger than our planet disappearing in an instant ..then you would probably fall to your knees in absolute terror!"

"So what? Because I'm terrified, it doesn't negate the science!" Peter said.

"I'm not negating the science, I'm just reminding you of their utterly alien nature – they deserve more than blasé tones!"

"Lighten up buddy"

"Consider your garage as a solid block of steel and then imagine it disappearing in a split second? I don't think so. And you ask how such worlds could escape the rapier sharp blade of science and remain undetected. It's really very simple. Science has unwittingly sealed itself into the arena of channel 458 through its very method."

"Oh really? Please explain."

"At its core, science assumes there is an external reality called 'the environment' which starts at those particles that are way way smaller than atoms and balloons out into the incalculably stupendous universe we observe. Science imagines the human brain as a sort of witness to this world. An infinitely complex sponge which absorbs and correlates information from the outside (the environment) and with supreme analytical method, makes sense of it."

"Well are you going to dispute that?" Peter said

"You know I bloody am! The earth and the universe in which it exists, are to science, rock solid observations, as is the human brain which observes them. Everyone knows the world is enduring whether we are here or not. We live and die and get buried in this world and it carries on rolling round the sun. Yes? Well I'm sorry to disappoint you but this is not true…..or perhaps more accurately it is true and untrue, depending on your perspective."

"Oh yeah?"

"No, the world that we live and die in does not exist independently of us and never has. This has been its great illusion and our great folly. Our world is generated by our mind tools in collaboration with channel 458 and we are tuned to it. The real world has an existence way way beyond our comprehension."

"Okay …what about the animals, the birds, the insects Adam …did you forget about them? They seem to agree that the world is real and has an objective existence!"

"Why, because a beetle falls off the edge of a table? Or a bird flies into a window? Like I was saying, all the living creatures we can perceive are tuned to the same channel as us ...are the same frequency as us… and therefore act out their lives within the walls of 458. The solidity of their world is the same as ours - in other words they inhabit the same perceptual reality or energetic landscape that we do but with much less clarity. But that clarity or extreme focus that we have on 458 comes at a price. It allows us little or no peripheral reception. A dog with its weaker focus may well have the advantage of a broader spectrum of reception. Yes to other channels. Hence the dog's legendary ability to sense 'ghosts' or navigate its way home over hundreds of miles of unknown territory."

"Here we go." Peter muttered.

"Our MT's don't normally have the 'vocabulary' to build ghosts, or more likely it's evolved out of us - although they would no longer *be ghosts* if we did, because they'd be a bona fide phenomenon in 458. That's because ghosts are irrelevant to our existence being entities in a foreign channel and therefore unnecessary clutter. The MT has simply screened them out."

"This is all so crazy Adam!"

Adam looked at Peter for a few moments and then said, "Pete, take a good hard look at me."

"What?"

"You heard ...take a look at me!" Peter lifted his eyes. "Okay…. you have to realise that you're now looking at the language of your mind tool. You're not seeing the real me. You're seeing your version of me. The person in front of you is just part of the symbol system of homo sapiens. A creation of the human Ape's mind tool. And in the same way, *you are my creation*. We seem real don't we! Don't be fooled. The truth is Peter we don't really know what we are. We understand ourselves through the language of the mind tool. Oh sure, medical science has stripped back the body and burrowed into its most secret interstices, but guess what? These discoveries are all the products of the mind tool. The alterations of the body 'as we understand it' will comply with our understanding of the machine as volunteered by our perception. *We all share the same one!* So it's all beautifully validated. Do you get it now?"

"I actually don't know what to say my old friend," Peter said looking at Adam with some concern.

"And were I to die right now…."

"Please!"

"You'd be left with my body, yes? But here's the thing. This body that you're observing, as we just said, is merely the language or production of the mind tool. It is not a real thing! In other words it's part of the energy of sensorium 458. A mind sculpture, now inert and devoid of life.

But is there something else about us as living beings which is *beyond* the ability of the mind tool to manifest? Outside its script? Now there's the mystery. As we die, the energy fields which hold us in 458 eventually surrender to the pressure of chaos or entropy and force us to dissolve out of this channel leaving only the energy husk behind – which then carries on disintegrating into chaos. But is there a dimension to us which is not the frequency of 458 and so is invisible to the MT and which persists? There's the mystery!"

"Are you desperate for an after life, Adam?"

"That's a whole different question. If you're thinking that I can't bear the idea of the finality of death and that's what's driving me to speak this way … well …you're wrong. More than anything I think what drives me to think this way is that I've worked out that the world we live our lives in is not actually real. That simple understanding means all bets are off! I mean come on. Once you've understood that even the hardest thing in the world is made of our mind language, then it's quite a shocking revelation!"

"I just can't get my head around this idea!"

"If you can only understand that granite for instance is generated in the vocabulary of the mind tool, then you're on the way to unlocking a profound secret."

"But granite is not made by the mind tool, Adam, it really is out there!"

"The energy that 'granite' represents is out there …sure…but the quality of hardness is a pure example of mind. You're only contesting this because you know how a hard object impacts your body in the real world. You just cannot stomach the absurdity of the concept because all your life, you've been under the impression that hardness is a quality outside you... existing in the objects of the world. It simply has to be real. But Peter, it is not! When you're not there to generate it, believe me, hardness ceases to exist! It's just a unicorn!"

Peter rolled on the rug convulsing whilst pulling his knees to his chest.

"You mock me …and don't think I don't know it!" Adam said light heartedly, looking at the bunched up figure of his friend.

Peter snorted. "You're right, I am laughing at you …but …but not in a nasty way. It's just this conversation is a bit of a mind bender and although I know you're speaking utter rubbish I'm sort of quite enjoying it. It's like I've been having a session with the craziest shrink known to man, who's convinced himself that his insane thoughts are valid!"

"Lordy, I feel so much better now," Adam said smiling, "Sort of lighter and unburdened."

"So my being your patient worked did it?" Peter said with a glint in his eye.

"Oh yes …perfectly. My mind is crystal clear now. The fog has lifted."

"Really?"

"Yes. Oh yes. Verbalising my thoughts has really made it all so much better. You'll understand what I mean one day Peter. Hopefully not at the very last moment. That would be sad."

"Are you going dark on me again, mate?"

Adam gazed out over the sea. "No ..not at all."

Peter sat up and crunched into an apple whilst they sat in silence for a minute or two, listening to the surf.

"You know … I *was* quite interested about the idea of insects having a different view of the world," he said presently, "you said they've got mind tools but with a different …er …"

"Vocabulary?"

"Yes. That's it. I've often wondered what insects actually experience. You know, what their world is like. Hell, they give me the creeps to be honest. I mean does a fly experience terrible panic as the spider advances toward it in the web?"

"I don't think so …I really hope not! Look, we agreed that although we're all living in the same energetic landscape, there will be a difference in the vocabulary of the various mind tools. In fact they'll more than likely live in entirely separate symbol systems. In other words their reality 'language' is unique to them; especially when we move away from mammals and enter other animal worlds as you said, like insects."

"Are insects animals?"

"Look let's not be pedantic. Perhaps I should've said creatures."

Peter picked up his phone. "No! You're right," he said after a few moments.

"Ok. The point is that we as humans believe that the world we live in, is *the* world. It is not - it is nothing but a highly evolved model. When you die Peter, you're not saying goodbye to a real world ….you're exiting an entirely false world. A world of symbols. The language of the Mind Tool."

"Yes, Sir!"

"Our perceptual world (which in effect is our real world) is so well defined by our superior cognition that we understand and manipulate it and enhance our survivability within its walls more so than any other living creature. But ironically it has also reached its peak in terms of its ability to deceive. They go hand in hand."

"Agreed!"

"Don't be sarcastic! On the grand stage our clarity is our enemy. We as a species are equipped with the strongest receiver if you like and because of it's power, our receiver (our mind), has made our world suffocatingly real - so real in fact that we operate within its dictums with supreme efficiency and have created mind blowing technology as a result. And this gives ever greater credence to our world as something that has an enduring objective reality. A reality which scientists have analysed, understood, documented and filed."

"Stooooop!"

"But there's a problem. The more we learn and the more we progress, the blinder we've become. Why? Because science is continuously ignoring something critical. It assumes the objects of its observations have the qualities it observes them as having whether they are being observed or not. In other words it tries to write itself out of the script."

"Well it was you who just said that all we ever perceive is our own mind anyway!" Peter said "therefore these 'objects of observation' as you call them may …may simply cease to be if they weren't being observed!"

"Exactly! Because all we ever perceive is the interface between our mind tool and channel 458. And this has its own rules of play, its own distinctive signature, its own filters. It arranges things in a very particular way and it chooses *what* to make available to us.

And what it chooses to make available to us is intrinsically related to our structure as beings and our survival and continuation as those beings. But of course we have no way of knowing this because we're embedded within it. To put it crassly Peter, if every time you looked at a lion it appeared as a parrot you would assume a lion was a parrot. You would be ignorant of the existence of a lion."

"Help me, help me!" Peter groaned theatrically rolling on the rug, "lions…parrots…what fucking next?"

"The real problem is the scientific method's addiction to the Mind and Matter issue, subjective and objective, - the ghost in the machine. We need to realise that there no such thing as matter as we understand it. That's an illusion. Mind and matter, objective and subjective are shades of the same colour and it's called Sensorium 458."

"Ok then, how the hell are we managing to progress, to manipulate the environment to create machines?" Peter interrupted lazily, " How can you trash the whole bank of human knowledge as illusion when evidence of our success as a species is all around us?"

"I thought I'd answered that. I'm not trashing anything Peter. I love all things scientific as you know. I love the achievements of Nasa. I'm just trying to make it clear that matter as we perceive it, including my car, is a feature of our perception and therefore is mind or awareness. That's what we term 'real'; it's just that 'real' isn't quite what we thought it was. The origins of that perception is energy of a certain frequency which we can manipulate to create the world we live in.

The laws of physics, chemistry, etc that have been so painstakingly revealed in the modern era reflect the exquisite fine tuning of our world and this in turn reflects the complexity of the relationship between us as beings and channel 458 . But they will not be absolute laws. On a wider perspective their efficacy will fail because their origins are strictly limited to the channel on which they were developed."

"Well I can't see a problem there because we *are* human beings and we're not about to morph into something else are we?" Peter said, chucking a pebble over the cliff.

"That may be so," Adam continued, "but I think you're wrong. Like we were just discussing, there are already evolved human beings amongst us. People who have powers and abilities outside our understanding of reality.

Powers we would label as impossible. And why don't they show up?"

"Well exactly!" Peter said leaning forward eagerly.

"I am not going to answer that, Peter, because it is so bloody obvious!"

"It is? Why? We need proof?"

Adam rolled his eyes.

"Proof?! I can only tell you there are millions out there who watched or heard that Neil Armstrong had landed on the moon in 1969. They knew that rocket technology existed back in the early forties because of the V2 weapons used by the Germans in the Second World war. But despite this they still choose to listen to the insidious rantings of the conspiracy theorists who proclaimed it was a studio creation. If you have people doubting an event

in which thousands of engineers were involved …which was tracked from start to finish, and which we had proven if shaky technology to accomplish, what hope have we to convince people? There will always be doubters."

"Okay…I'm with you on this one."

"Exactly! So there is no such thing as concrete proof Peter. So why would an evolved being subject him or herself to that circus? What would be the gain? It would all be deemed an elaborate trick! He'd be the focus of 3 billion people all hounding him to spill the beans."

Peter bounced a pebble on the palm of his hand for a few moments whilst gazing at the horizon.

"Mmmm …guess you have a point," he said eventually.

"Millions believe in the existence of Jesus Christ …that's until he actually shows up. Then you only have to imagine what would happen. The conspiracy theorists would go into hyper drive and even when he split the Pacific in two, they'd claim the whole world had been under mass hypnosis or it was a gargantuan optical illusion. So guess what. He isn't going to show up riding a chariot of fire".

"Do you believe he existed, Adam …seriously?" Peter said

"Of course he did," Adam said, looking out over the sea. "He was a very powerful being on a mission …but somewhat dressed down for his visit into Sensorium 458, which if you think about it is the only way to come."

"Mmmnn," Peter said thoughtfully.

"Anyway …let's not lose track …we were talking about …erm.."

"Scientists."

"Yes that's right. Surely you understand that most scientists are inflexible? They have a supreme advantage! The scientific method is so successful that we treat it with reverence. Understandably. Its proclamations are not to be refuted. Well …conspiracy theorists excluded.

For instance we know that if a man can copy a drawing in a sealed envelope without ever having seen that drawing, then he must be a trickster or lucky. And when he does it again and again and again then he's just a plain trickster ….because we think we know what can and cannot occur. If a man says he saw a ghost in his kitchen, science tells him politely he's mistaken. And science is like this. A sort of Sherlock Holmes figure, who always pierces through silly fancies and swats them away with his titanium cane."

"You're beginning to sound like a rather poetic preacher, Adam!" Peter said lazily watching the clouds

"Probably! You see when science is confronted by something which doesn't fit into its understanding of reality, it unleashes a barrage of accusations. When Geller bent spoons, the sober sentinels of 458 came up with a chemical which had the same effect and accused him of using it. When he quite obviously didn't have this chemical they accused him of sleight of hand! When that didn't wash, they just went back to calling him a trickster …a very clever trickster. Are you falling asleep Peter?" Adam asked suddenly.

"Er ...no ...don't think so."

"Okay ...so going back to what we were talking about" Adam resumed, "the long struggle of humanity to understand the world and the rules by which it operates is a stupendous achievement. It gives us a feeling of great power and security and many scientists believe there will come a day when we can close the book of knowledge, crack open a bottle of champagne and utter the words.. *'We did it!'* But this will never happen because this world is not existing independently of us - scientific certainty is like the end of the rainbow…always moving out of reach. We are living in a very particular interface between an unfathomable reality and our mind tools."

"As you keep saying."

"You have to realize, 458 is laying our table all of our lives and as such, it has a suffocating hold over us. But it will always leak around its outer boundaries where phenomena are beyond its jurisdiction… like telekinesis, poltergeists, ghosts, crop circles, unexplained sightings etc. Science loves to place as many of these inexplicable oddities into a neat little box called 'subjective experiences' such as dreams, hallucinations, wishful thinking, momentary lapses of consciousness, fakes etc. because it suits its overriding desire for certainty. And certainty is the nectar of science.

In fact the bulk of scientists unwittingly man the watchtowers of Sensorium 458. They're just too busy enjoying the dish that 458 is serving up, to question the origins of its cooking. Why would they? They're making our material lives better and they've rightfully earned tremendous respect for it".

"Are you being patronizing?"

"But there can be a problem here. Where science sees phenomena which don't fit in with its understanding of how reality works, it actually begins to discredit that phenomena. In other words science begins to bend reality to fit the matrix it's superimposed over it. Much of this has to do with reputation and credibility. It takes a brave scientist to preach his belief in something unorthodox, or worse, something that has the whiff of 'new age hippy' about it; and this makes scientists very cautious. They don't want accusations from their peers of going soft in the head and losing their grants, in the same way that pilots tend to keep quiet about strange happenings during flight – they want to keep their jobs and the approval of their peers."

"Okay."

"People only have to hear the word 'hoax' or 'fake' to recoil in terror. The indignity of being fooled lies very close to the hearts of everyone. And because we live under the dazzling glare of 458 it only takes someone to 'mimic' a real paranormal phenomena for that phenomena and all its ilk to be treated with derision and ridicule. But remember, throughout history visionaries have been discredited by conventional wisdom because they dared to challenge an orthodox view."

"So you're a visionary now are you, Adam?!" Peter said rolling his eyes in mock laughter. Adam ignored him and started counting on his fingers.

"The earth going round the sun…the earth as round not flat…man from ape …Einstein proving Newtonian physics was flawed. We need to realize that even our most coveted scientific laws may one day have to be relinquished. Weird phenomena like black holes which cannot be placed in the 'subjective experience' or 'theory only' box, because their existence is undisputed, sit uncomfortably in our world.

"Aha! All these things you mention have been proved by science! By the scientific method."

"So? Peter you seem to think I hate science or the scientific method! But I do not! Science is a system of trial and error in the light of evidence as presented in sensorium 458. It's built on logic and it's got the human race to where it is now. My gripe with science is when it becomes a tyrant and brushes aside evidence that runs contrary to its tenets. You know …that whole thing of, "you didn't see 'that' because 'that' is impossible!"

"Look I know what you're saying, but there's so much twaddle out there that science is naturally conservative in what it 'allows' to be taken seriously." Peter said, sighing.

"That's exactly what I'm referring to! What it 'allows' to be taken seriously. That's where scientists can become dictators. Not all by any means but a hell of a lot!"

"But you accept that there is a lot of misinformation out there?"

"Of course! People will believe the most ludicrous things. I'm just making the point that rock solid theories of today can easily be discredited tomorrow. The idea that our reality is one of many and that they exist all around us, is still a way off main stream thinking but already there are some highly reputable scientists who think this is probably the truth."

"Some people would say that is a prime example of a ludicrous suggestion, mate!"

Adam laughed and slapped his knee. "True! It does come over as barking mad, I completely get it. It's loony bin material. But that's reality for you. Stranger than we can imagine."

"Well at least you admit it!" Peter said smiling.

"Since we're in the mood I'll add some more craziness now. As I said earlier, I think this world is inhabited by many beings we haven't recognised as yet. And do you know how we're going to discover them?"

"Go on."

"Through Machine Intelligence!"

"What the hell are you talking about?

"M.I. will reveal those entities that humans throughout history have encountered but which science scoffs at. Why? Because MI will have eyes everywhere twenty-four seven, recording what it's observing. No chance for doubters to say 'it didn't happen'! Added to that it'll be armed with a vast range of sensors well beyond those of humans. It'll detect in the Ultra Violet, the Infrared, Xray and so on.

"But we have all these sensors already!"

"Yes we do …with human operators. But here's the point. These sensors will soon be distributed amongst machines on a huge scale, including robots, and they'll be networking with each other continuously and on a scale we can only imagine. Humans and their prejudices will be out of the loop. What humans shy away from because it comes under the label of paranormal will be replaced by machine intelligence which doesn't have preconceived ideas about what is 'supposedly' normal and what is abnormal. To machines, these phenomena are just what they seem to be. Phenomena. There will be no preconceptions as to what the world is 'supposed' to be.

Weird, unexplained phenomena which we are currently witnessing in isolated pockets around the world and which are often recorded by people in the armed forces, in civil aviation, in space agencies will suddenly explode into the open. It won't just be a couple of guys on a flight or a dozen people in a city park who'll be witnessing these things, they'll be recorded on a huge scale by machine intelligence."

"But we have cctv all over the place as it is!"

"Sure, and if you care to look, they're already revealing some very peculiar stuff. Stuff that the 'sensible' wooden people claim are glitches in the system. But once machine intelligence starts to network on a huge scale, like I think it will before long, then you'll see some mind blowing revelations!"

"You think so?"

"Yes I do! Machines don't have self doubt, suspicions, preconceptions, embarrassment, grants to secure or career fear to contend with. They'll tell it as it is. And they'll establish that so called *paranormal activity* is super abundant. Their synthetic neural networks will be *talking* to each other, gathering information and analysing. They'll discover patterns, concentrations, distinguishing characteristics etc. They'll gradually establish the existence of a hidden world. They'll inform us, (their human masters), that we have *company*."

"I suppose so. But there's something that confuses me. You've made a big deal of saying that all phenomena are subjective. Creations of our mind tools. So how do the phenomena that machines are observing, fit in with that?"

"Machines, as we said earlier, are sort of enhanced versions of ourselves. They have the same 'mind' as us. They live in the same world as us or more accurately the same energetic landscape. Their observations or brains are inevitably tuned to the world of sensorium 458. So the phenomena that they encounter will never reveal the true essence of what they're observing, only their interpretations of those phenomena. Just like us. But let's not forget at this stage they're still unconscious machines which are merely presenting their findings to us their masters. Their advantage over us will be in their ability to correlate and analyse the information at their disposal on a huge scale, in very fine detail and at incredible speed; not in their ability to dig deeper into the origins or sources of those phenomena. Am I making sense?"

"I think so. But to go along with this … I've just had a thought!"

"Go on."

"Will machines have mind tools?"

Adam stared hard at him for a moment or two. "Hey … we're having some incisive thinking here, my old buddy!" he said grinning. "That is a very good question …*does a machine inhabit a mind language like a human?* Well the human is consciously living in its mind language. It's called the world. Like we were saying earlier, the human is continuously mistaking its symbol system for a bona fide reality. That's the power of the mind tool. But as to whether machines will eventually inhabit a symbol system like a human is …is a bloody difficult question! Certainly not yet. They're not nearly advanced enough. But quantum computing I suppose may make it all possible!"

"For instance, could machines inhabit the same world as us?"

"They will inhabit the same *energetic landscape* that we do because we've built them in the world of Sensorium 458 and with the energy or matter of 458. They will be compelled to understand the world as we understand it. First on that list of course is that quality we call hardness. That's something machines are compelled to have as a cardinal symbol just like us or they will 'die.'"

"You mean the machine will have to comply to the constraints of the material world?"

"Of course! We take it for granted because of our utter conviction in the external presence of our world. The robots, like us, are held within the walls of Sensorium 458 and will be obligated to obey its rules. You have to realise, machines, and all the components of those machines are artifacts of our reality. And our reality is governed by our own energetic configuration. As I've said before, our reality is built on an interaction between us as sentient beings and the mysterious environment in which we live - an *interaction, an interface between our energy and the energies that surround us!* Its appearance as a bona fide 'place' is illusion!"

"Are you going whacky on me again?"

"Look, it may sound whacky but that's purely down to unfamiliarity. Science has found countless whacky things in its quest for truth …it just goes with the territory."

"So to go back, we and the machines will inhabit the same material world?"

"Yes! Otherwise drones would be splintering into pieces when they flew into mountains. But they won't 'live' in the same symbol system as humans. Why would they? Ours has developed over millennia and it's not something we can hand over! I cannot find your perception of me and extract it from your head, Peter! I can extract the facilitator of that perception but not the perception itself. As we already said, the mind tools of the different species on this planet are probably unique to them but they all inhabit the same energetic landscape as us. Although …not exclusively. Like we said, animals may have more fluidity on this front and have a mind tool that utilises other channels."

"But, are you saying a machine could feel this world, like us? Within its own symbol system of course - like we do as we sit up here on this clifftop with the sunlight, the grass, the views?"

"But Peter, we've already established that this clifftop, views etcetera are part of the human mind tool's rendition of reality. It isn't actually real. In the sense of it being here if we're not here to generate it!"

"I understand that … what I meant was …was…oh forget it!"

Adam looked at him. "What you started to assume was that this clifftop is real and can the robot sense it like us! I get it. It's almost impossible to overcome the certainty of the external world as a 'place'! So the robot will not experience the same 'place' as us but it will certainly sense the physical parameters of this environment … because we've designed its brain to think like this."

"Well we couldn't design its brain any other way could we? We can hardly introduce it to the 'other channels'!"

"Yes! You're absolutely right. It has to 'live' in our world or it would be a very strange robot indeed. In fact a bloody scary robot!" Adam said throwing his head back and laughing. "It would become like those UFO's. Capable of extraordinary acts outside its designers comprehension."

Peter looked at his friend. "This conversation is truly crazy mate … I mean what the fuck are we talking about here?"

"It's good mind flexing isn't it?" Adam said with a cheeky glint in his eye.

"Ok …ok … let's go back to what we were talking about."

"Which was?"

"About robots experiencing the same world as us."

"But we decided that they could not experience our actual world because that is a human code if you like. They would live in their own 'code' or symbol system but one that is using the same energetic landscape as their template."

"Right, I've got it. I was just interested to know whether the robot will, as it were, 'live' in its symbol system like humans do."

"Well, we're now talking about that elusive quality called 'consciousness'. I would think that robots will become incredibly smart within the orbit of their understanding of the world. An understanding given to them by us. And which may self evolve beyond us! But I think machine learning, however smart …will remain cold; trapped in the universe we have instructed them to cognise. In effect the universe as described by the human mind tool. As I mentioned, hardness or density will be one of the corner stones of a robots' understanding of its environment in tandem with the cognition of space, distance, temperature and of course, time. These are presented to them by us, according to our own understanding of reality. We're passing on to them our own 'Mind Tools' if you like, which will sense what we sense but with incredible speed and hyper integration."

"And sometimes beyond what we can sense, as we were just saying."

"Correct, but without the 'other' component that humans have."

"What do you mean?"

"Humans seem to have a double mind. They can watch themselves. Analyse themselves. See the robot in themselves outside of the auspices of the Mind Tool *as it fixates on channel 458. I* suppose you could say humans have *'will'*… an ability to challenge themselves and their instincts.

It's like a component that sits outside of the machine if you like. I'm not sure that all humans have this facility or perhaps it's just dormant but I guess all humans have access to a deeper consciousness that will not be present in robots. It's almost like we're able to sneak around the MT and observe it's machinations from the outside."

"You think so?"

"Yes. Remember, my old friend, the human Mind tool is not a servant of channel 458 exclusively. The Mind tool has many masters - *The 'other' channels*. Look, I'm reluctant to say this again because it sounds what you'd call new age rubbish, but I'm going to anyway. Like I mentioned earlier, regarding our 'advanced settings', there are humans who've learnt how to manipulate the mind tool to focus on other channels. Yes, humans who are psychologically very advanced. This isn't something you can learn from a book. This requires years of teaching and instruction. And the reason for this is that it is a very dangerous and precarious path to take. It's been said that no sensible human would ever enter those realms of their own volition. They have to be tricked into it."

"But you just said that there are places to learn this type of thing?"

"I didn't mention 'places' to learn. Like a school where you pay your dues and sit in a classroom! I mentioned teaching and instruction. But the point I'm making is that you sign up to this 'teaching' without realising it. You become involved in it in an oblique fashion and never really understand where it's all leading until it's too late to turn back."

"It all sounds bloody sinister mate ... like those cults in the states where people become brainwashed and can't escape."

"Yes I imagine it does. I assure you though, this has nothing to do with surrendering your material possessions to some fraudulent guru who plies you with drugs and gets his rocks off at every opportunity. It's something different entirely. This is knowledge that leads to the breaking down of the walls of this world. Literally getting behind the mirage of the mind tool and subverting its control."

"My God."

"Without instruction in this enterprise we're crushed like bugs. The impact of this type of knowledge is truly awesome. As I said earlier, the mind tool is not only a dictator, it's also a protector. It shields us from the full blast of reality. But in order to do this, it has to utilise a code. In the case of humans, it's Sensorium 458. If we were tuned to … say ..channel 297, we'd be living in Sensorium 297 - an entirely different energetic landscape and one which would require we alter our energetic configuration - if we were to try and function in that reality. This is how it works. We live our lives in a false arena as created by our protector and ruler."

"So to go back, this 'other' component is … is something humans have but machines cannot have?"

"Yes. Most humans don't even know this component exists within them. It's a deeper consciousness which in most of us is dormant."

"Ok."

"Anyway, back to robots, humans naturally think that once we've reached a certain stage of engineering and technology, robots will out match them on every front. *Which they will!* Because the robot will be the embodiment (in superior form) of the machine we mistakenly think we are. But what we think we are is itself a creation of the mind tool. And the mind tool's presentation to ourselves of ourselves is a construct of the relationship between our energy configuration and channel 458."

"God almighty!" Peter said looking up to the sky.

"I know this sounds heavy but …!"

"Heavy? This is like wading through treacle!"

"Sorry, just let me finish. The point I'm trying to make is that our understanding of what we actually are, can evolve away from the dictatorial constraints of channel 458 and lead us out of this prison."

"So then we become channel hoppers?!"

"If you want to be flippant, yes!"

"Ok …ok. But do you think robots will actually become a threat to the human race?"

"I really don't know. Hawking seemed to think so. Perhaps in the future, robots really will develop malign feelings towards humans and try to assume power. That's when those advanced humans will have to manipulate the mind tool's focus, and navigate out of Sensorium 458 with some urgency! But it won't be either of us that's for sure!"

"Really?"

"C'mon. You think it's all a load of bunkum, and I don't have what it takes!"

"Guess so!"

"But going back to machine intelligence, I believe the whole UFO scene is going to be opened up as well. Those strange sightings will suddenly be under the surveillance and more importantly the *analysis* of machine intelligence on an ever increasing scale. Information currently under lock and key in government buildings will be looked at with fresh eyes."

"So you believe they exist?"

"UFO's? Or as they're called theses days, UAP's - unidentified aerial phenomena? Of course they bloody exist! Only someone who's completely disinterested or someone who's mind is bolted shut, could refute their existence! They've been observed throughout the world, Peter, for decades. Various US aircraft carriers have had them buzzing around them for months, with video footage as proof."

"Oh …I've never really taken them seriously."

"Well you'd better. They're definitely here. The problem is that the powers that be, especially the US, tries to hide evidence of them."

"Why?"

"Because they recognise that these things are way way ahead of them in a technical sense and that the general public might freak if they knew too much. Look, it's not so difficult to understand.

The US is the top military on the planet. They don't want it to be general knowledge that they've got UAP's flying around their ships and nuclear facilities which they can't do a thing about! It's humiliating and downright terrifying at the same time."

"But are they seriously doing this?"

"Yes, Peter, they are. And nobody can do anything to stop them! I think they've got to be extra-terrestrial vehicles piloted by living entities or autonomous extra-terrestrial vehicles … managed remotely. Or even vehicles which are trans channel. In other words they can appear in channel 458 and interact with the 'energy' of our channel but which are not subject to the physical laws of our channel."

"Oh Jesus you're off again!"

"Look, you need to stop thinking of trans channel as something mystical. It's cold hard physics Peter. Physics as yet not understood and therefore dismissed as poppycock!"

"Don't make me laugh! What do you know about physics?"

"Enough to understand that today's scientific boundaries will seem primitive to humans of the future."

"Okay. So what makes other 'channels' a necessity for UFO's?"

"I didn't say a 'necessity'. I said it might be a possible explanation for their behaviour. Because they do things which seem impossible. They don't really fly - for a start they don't have wings! They dart from one position to another in the blink of an eye and execute turns and manoeuvres which are way beyond the abilities of our most advanced fighters. They show utter contempt for inertia and the laws of gravity. They make no sound. They appear to have no mass and yet are capable of affecting solid objects in our world. Sometimes they're tracked on radar which implies they are solid. The bottom line is that they do not appear to be made of the same energy frequency as us."

"I've heard it said that these vehicles are merely cutting edge 'black ops' created by the US military!"

"Believe that if you want, but they would say that wouldn't they! Of course they want to take ownership of these things to try and scare their enemies by implying they have incredibly advanced tech. But I don't believe it. The technological leap that these things represent is outside the scope of human engineering. Why would you spend decades designing and making the F-35 fighter jet when tucked up your sleeve you have this far superior tech? It's illogical."

"I hadn't realised you were so into this stuff Adam!"

"I keep my eyes open, my old friend, that's all. UFO's are like the rocks falling from the sky 200 years ago. Science is pretending they don't exist. But shortly it'll have to. I mean how stupid do we have to be to imagine that we're alone in this vast universe? It's a joke! We're jumping up and down with excitement at the mere thought of an extra-terrestrial microbe when we've got these 'things' darting around our warships. We need to get with the bloody programme. The time for our species to groom itself in the mirror over its technical achievements is over. We need to wake up!"

"You seem very sure! Many scientists say that interstellar travel is impossible because of the constraints of light speed. That these things cannot be here for that very reason."

"There's so much evidence, Peter! Soo much. And what you've just said is exactly what I mentioned earlier about the tyranny of science. *'Light speed is the ultimate speed,' they proclaim 'and therefore sightings of extra-terrestrial vehicles must be false! Because they could never get here. We in our ultimate wisdom make this proclamation with complete certainty!'*

Well they are here, so rather than try to discredit the facts, show some humility and accept that we still don't know what possibilities lie outside our comprehension. The arrogance and small mindedness is breath-taking!"

"True. Homo Sapiens is smart but not quite as smart as it thinks it is!"

Adam gave him a high five. "Thank you! That's the best thing you've said all day, my old friend!"

Peter smiled and chucked another pebble over the cliff.

"The industrial revolution happened some two hundred and fifty years ago and we're now the brightest kids on the block, I'll have you know!" Adam continued.

"No, the galaxy if you please!" Peter added.

"Oh come on ...shall we go the whole hog and just say it ...the Universe!"

"Yeah ...and there's none as clever as us. There may be a few microbes in a filthy puddle on some far away planet but come on ...we're at the apex of God's creations," Peter said grinning.

"Too right. We're his favoured ones, blessed to only look down on the lesser intelligences, whatever muck they inhabit," Adam said, trying to keep a straight face.

"Agreed. Look at our missiles and spaceships and formulae!" Peter added with a final flourish.

Adam burst out laughing and stuck his head into the grass, convulsing. After a few moments he gathered himself and sat up again.

"I know I harp on about this, but we've got to always keep in mind that all phenomena are part of the vocabulary of our mind tools, including UAV's. They're a perception!

Our mind's rendition of 'something else'. But what that 'something else' is in this instance is truly baffling. It's almost like we're seeing something outside the arena of Sensorium 458. But that shouldn't happen."

"Why?"

"We've been through this Peter …the whole thing of channels and their frequencies. Another thing is we shouldn't bunch these UAV's into the same basket. There might be worlds of difference between them in terms of technical prowess and advancement and for all we know there's a pecking order up there that we're completely blind to. Perhaps the hyper advanced ones snap up ones further down the food chain and take 'em home for research. The earth could be a sort of watering hole where the king predators lurk in the bushes, so to speak, in order to catch the lesser predators."

Peter burst out laughing. "What, sort of like wildlife observers kidnapping each other! You really are a crazy kind of guy, Adam! I mean Jesus …what next!"

"I try to please."

"I once read that submarines also encounter weird phenomena."

"Absolutely. They've got a name! USO's. Unidentified Submerged Objects!"

"Really?"

"Yes. There's a whole bundle of evidence of these things too. They've been spotted coming out of the sea, into the sea, travelling at ridiculous speeds down there. Also various lakes. Again, the 'sensible' people will say it's just glitches on the systems. They always want to explain away things that disturb their world view."

"I guess so. Military types can have very wooden minds!"

"Yeah. They cling to explanations that are more unlikely than the events themselves! It's a bloody joke! Do you know, they were analysing the performance of a new torpedo off the east US coast in the nineties which meant it had to be retrieved by helicopter after it had done its trial. But before they could get an anchor on it, something came up from the depths and took the torpedo under their very noses!"

"Russian or Chinese sub?"

"They said categorically, *it was not a sub.* They were naval guys, and knew what they were talking about."

Peter blew some air through his lips and shrugged.

"There's all this stuff out there Peter…if you bother to look."

"Ok."

"Another thing that struck me was whether an alien, high tech species had actually managed to embed itself in the US military industrial complex already! It seems there are so many clandestine components, so many hidden tentacles in that organisation that no one person is cognizant of what's going on in its entirety."

"I can't believe that! But there again isn't that the theme of the Terminator?"

"Probably. It does seem far fetched I agree, but it sort of makes sense! Infiltrate the central nervous system of another civilisation rather than fly out of the clouds with a 'Greetings' banner floating from your saucer!"

Peter snorted. "Possible, I suppose."

"You only have to listen to what some of the naval pilots say, to know that they've given serious thought as to whether their strange encounters are products of their own military."

"That's what I was saying."

"Absolutely. But it didn't make sense. Even so, they still had that suspicion to start with which just goes to show the degree to which nobody quite knows what everyone else is up to. And that leaves a crack in the system doesn't it?"

"I guess it might do," Peter said.

"People ask why these 'visitors' don't make formal contact. But it's obvious why they don't. I'd even go so far as to say that space faring civilisations have a code of conduct. Never intrude on a nascent technical society in an overt manner. Especially one which has learnt how to build nuclear weapons. Why? Because, more than likely, it'll cause all out war. Think about it. Every civilisation that evolves in our universe starts off by believing it's alone. That it's special!"

"You think?"

"Yes I do!"

"But why would the landing of an alien craft bring about war?"

"Because there would be great suspicion and distrust. Do you really think that because Fox news broadcast that extra-terrestrials had landed on the outskirts of Washington, that everyone is going to miraculously believe it? No bloody way! It would be seen as a ruse to gain some sort of control. And if 'extra-terrestrials' were shown on tv walking out of their spacecraft it may well be understood as the unveiling of a new weapon by the US – a new breed of robot programmed to spread across the planet to 'help' humans."

"Well I'm not sure about that. Though it must be said religion can very easily escalate into war."

"Over the millennia, religions have formed which have very specific ideas of what higher beings are. They wouldn't take kindly to beings that have come from the sky and don't fit their preconceptions. They'd call it a fraud. A blasphemy against God. Then all hell would break loose and certain 'leaders' would be propelled by their supporters into taking drastic action. Look, there are some pretty fanatical leaders who have very powerful weapons at their disposal and who would probably be happy to unleash them to defend the dignity of their God."

"But why would a space craft from another world be blasphemous?"

"Because it would not fit their religious beliefs. They'd suppose it was a diabolical trick cooked up by the great Satan to undermine them. That's presuming the 'alien craft' landed in the US! In fact that would probably be the very worst place they could land! It would be more prudent to land somewhere low key. New Zealand perhaps."

"Jesus!"

"Or worse it could be seen as some sort of collusion between the US and an alien race which is putting everyone at risk! So my belief is that they don't make formal contact for this reason. The irony is that they probably were in open contact with humans in the past, but now we're so technically capable of destroying ourselves, they don't risk it."

"So your concern is that they could cause war amongst humans, not that …."

"Humans could be a threat to them? Get real! We're a bunch of primitive apes to them. I would imagine they could dispatch us with the greatest ease. They also know that their appearance will cause unimaginable horror in the minds of humans and literally blow our circuits. Humans in the past were less self important, Peter. They knew their vulnerabilities and frailties and accepted it. They sought solace in the Gods. They would not have been so fazed by the appearance of beings from the sky – they already presumed they were there. Modern humans seem to think they're the smartest, technically brilliant creatures in the universe. We're a science driven society. We think we know what the world's about and in the West, particularly, God is an irrelevance and aliens highly unlikely. But I don't think even the most open minded, alien embracing humans can imagine what sort of creature might step out of one of those saucers! We're very sensitive to appearance. I mean even a different skin colour has us fighting. So what would happen if a fifteen foot tall 'ant' scuttled forward!"

"Oh Jesus don't! That sounds utterly hideous. And with superior intelligence … bloody terrifying!"

"Too right! And even in the west there'd be deep suspicion as to whether it was real or fake or a mutant cooked up in some laboratory."

"You're right there, mate. When you look at the endless conspiracy theories out there it seems that nobody believes anything these days. I mean look at you. You think that reality itself is a conspiracy!"

Adam burst out laughing. "Not really my old buddy. No one's purposefully misleading us! The reality we find ourselves in, should actually be seen as a superb instrument of navigation to help us live our lives. You could say it's a dictator but equally, a protector."

"Okaaaay."

Peter stared out over the sea for a few moments, thinking. He then looked at Adam. "There's one thing that puzzles me." he said.

"Go on."

"You were talking earlier about the universe as a creation of the mind tool and not a 'place' as such."

"Yes."

"But just now when we were discussing Ufo's you seemed to speak about the universe as though it is a **real place** after all and not just our personal creation. You mentioned space faring civilisations."

Adam paused whilst gathering his thoughts. "The universe is definitely a mind creation, but that doesn't mean that …how shall I put it …that its realness belongs exclusively to us or that we have ownership of it. It belongs to any being that is fixated to channel 458. So if I seemed to be talking about the universe as a 'real place' I'm merely describing it from the human standpoint and to any being fixed to 458. Be sure, there are plenty of beings who live in the same energetic landscape as us, some of whom will be much more technically advanced."

"Ok."

"So when it comes to Ufo's we have to think to ourselves that some of these visitors *are* from our own 'channel' and abide within the physical constraints of our universe, whereas others are visitors from other channels or 'alien universes' who are not bound by the physical constraints of our universe. Does this make sense?"

"Not really ... but carry on.".

"Which begs the question, have any '458' Ufo's ever reached here or are we actually only ever witnessing Ufo's from different channels?"

"And does the earth even exist from the perspective of these other channels?" Peter added.

Adam shrugged. "What the earth 'looks like' from the perspective of one of these channels is a mystery. Remember, once we're talking of beings appearing from different channels we're not speaking of coming from *somewhere else* in the human universe, like a planet so many light years away. We're speaking of beings who live in worlds, *right here* with us now!"

"That's just impossible to imagine!"

"Of course it is! Which is why most Ufologists assume, like you, that Ufos are craft from somewhere else in our universe with vastly superior technology to our own, including cloaking tech and propulsion systems. How else would they overcome the laws of physics that exist (or are supposed to exist) across the known universe?

"Very true."

"But if you go down the route of there being many different channels and therefore correspondingly different energetic landscapes and laws of physics, then these craft have plausibility. Don't you think it's odd that no one has ever found wreckage of these machines?"

"Well some people say they've found some, don't they?" Peter said.

"In very very few instances. And are these genuine? I think it's much more likely that they haven't found any 'hard evidence' because these craft are not built of matter as we recognise it. Or more to the point, matter that we can see and feel. In our world Peter, 'feeling' is believing. You'll often hear people say 'does it have solidity?' That's our absolute arbiter of 'reality'. If it's solid then it's real! But as we've been discussing, everything we perceive including solid objects are inhabitants of Sensorium 458 and have been created by the Mind Tool. You may have heard about the ghost rockets in Sweden in the 40's and then again much later like the 80's. Hundreds of people saw them crashing or submerging in lakes but no one never found a thing! The wooden minds said they were shooting stars or planets of course."

"Really ...sounds bizarre!"

"I guess certain people believe our current idea of reality is set in concrete and anything that challenges it must be plain wrong. It's really sad."

"And makes them look foolish."

"It's the ostrich effect!"

"Don't they often get radar fixes on these objects?" Peter said.

"Very good point," Adam said nodding his head, "They certainly do! They often seem to have mass or at least are real enough to return radar signals. But does that categorically mean that the matter they're constructed of, is 'matter' as we know it? In other words is it perhaps possible to get an electro magnetic return from an alien vehicle built of exotic matter?"

"Well that is a question and a half, Adam! Who bloody knows because I'll bet you science has no belief in 'alien' matter. But then, that's been the theme of our discussion hasn't it! Worlds that exist right before our eyes but unavailable to our senses because of their frequency!"

"You said it my old friend."

"Dear God, all these crazy things!" Peter said, gazing out over the sea. They sat in silence for a few moments listening to the low rhythm of distant surf. Then he stretched and glanced at his watch. "Well buddy, it's time to hammer some rocks! I need to give my poor brain a break! You've certainly got me thinking." He stood up and shouldered his rucksack. "Okay Bud, it's been an entertaining and rather unexpected conversation …but hey … I'm out of here. I'm going down to the beach to start looking for bones."

"Sure…good idea!" Adam said also getting up. "Go and wrap your arms around a boulder and whisper to it that you know it's real …. cos it's heavy and hard!"

Peter cackled loudly. "You bloody nutcase, Adam!" he said, as he descended the rickety wooden steps.

Adam watched him for a few moments and then cupped his hands over his mouth.

"Just remember Peter," he shouted after him, "our world is pure mind and always has been!"

But there was no answer – just the wind, which snatched his words and threw them skywards.

Epilogue

"So let's begin," the Praetorian (mind tool) said, with an air of concern. "You seem to harbour grievances."

I peered into the darkness but I couldn't see her amongst the flickering shadows away from the fire.

"I'm not saying I have grievances, I'm really asking for clarity, I suppose," I said, sounding unsure.

"Clarity? You mean you want an explanation?"

"Yes. I want to understand why you shield me from the truth."

"What do you guess the truth is?"

"I don't know! That's why I'm asking!"

"Ok, listen carefully. Firstly understand, it is I who build your world."

"What? That cannot be true!"

"Thank you! That is a tribute to my artistry. My task is to protect you …"

"From what? I don't want to be told lies to protect me."

"Please. Don't interrupt. You've asked me a question and I want to answer it. If you snap back at me when you've only heard a small portion of my answer, then it is you who is causing confusion!"

"My apologies."

"As I said, I am here to protect you. And myself to be honest. Your fate and mine are inextricably linked. When you're finished, so am I, and that scares me. In a strange way, you keep me alive, so … I've become quite attached to you."

I felt a shiver of uneasiness.

"I'm sorry to make you feel uncomfortable. Firstly, let's be clear - the world is not what you think. It is a place of extreme hostility and predation. Yes, you heard that correctly. You are being hunted from the moment you were born!"

"Really? By whom?"

"By entities and forces that you have no knowledge of …because of my intervention. You must realise, your energy is prized. My job is to allow you to navigate the road of life with as little hindrance as possible from extraneous forces that …that …." She broke off suddenly and I was left, looking around nervously into the dancing shadows.

Then I heard her voice again, more distant now, and agitated. "I must go now," she said, hurriedly, "we have company. I'm taking care of you as I always have."

I felt a rush of fear, then a jolt awoke me. The garden was bathed in evening sunlight. I yawned and glanced uneasily around me. It seemed I had fallen asleep on the bench under the mulberry tree...

Printed in Great Britain
by Amazon